CLIMATE CHANGE IN PERSON, COMMUNITY, AND PLANET

CLIMATE CHANGE

IN PERSON, COMMUNITY, AND PLANET

A Guide for Helping Professionals

David Derezotes, LCSW, Ph.D.
University of Utah

SAN DIEGO

Bassim Hamadeh, CEO and Publisher
Amy Smith, Associate Editorial Manager
Abbey Hastings, Senior Production Editor
Jess Estrella, Senior Graphic Designer
Kylie Bartolome, Licensing Specialist
Natalie Piccotti, Director of Marketing
Kassie Graves, Senior Vice President, Editorial
Alia Bales, Director, Project Editorial and Production

Printed in the United States of America.

Brief Contents

Detailed Contents

Land Acknowledgment

This book was written on land taken from the Ute, Dine' (Navajo), Paiute, Goshute, and Shoshone nations.

Preface

A book written today about the climate crises, designed for helping professionals, also needs to be one about inclusivity, consciousness, and connection.

This is a text that includes descriptions of the many ways in which humans are connected with the world and discusses how our conscious awareness of these connections can help inform the practice of helping professionals, as we engage, assess, and intervene with our clients, communities, and ecosystems daily.

What are some of the most important connections that we need to keep in mind?

First, we could start by asking about the fundamental connections human beings have with ourselves, other people, other living things, and the ecosystems that support all life. In what ways, for example, does the quality of the environment in which I live influence my psychological and physical health? To what extent does my access to clean air and water, healthy ecosystems, and a night sky unpolluted by artificial light influence my well-being? And how do our tendencies to surround ourselves with pet animals, potted plants, gardens, and outdoor views reflect our deeply instinctive need for what E. O. Wilson (1984) called "biophilia"?

In addition, we might explore how human beliefs and activities are connected with the climate crisis and specifically the extent to which humans have contributed to global warming, ecosystem degradation, and species extinctions. More specifically, for example, we can understand how the greenhouse gases we continue to release into our atmosphere have contributed to such environmental disasters as increasingly intense droughts, flooding, and fires. And if we have, in fact, largely created this climate crisis, how can we best reverse the damage and heal our ecosystems and communities?

We can also explore the extent to which human well-being and suffering is connected with our climate crisis. How do, for instance, extreme heat waves, air pollution, and water pollution affect the well-being of people as well as our economies? What does the recent discovery that tiny microplastic waste can now be found floating and lodged in the organs of our own bodies imply for the well-being of our children and grandchildren? How can professional helpers, working in such fields as social work, education, nursing, education, education criminal justice, human services, addictions, public health, and psychology, effectively include environmental factors in our assessments and interventions?

Surgeon General Vivek Murthy (U.S. Department of Health and Human Services, 2023) recently warned us of an epidemic of loneliness, depression, and anxiety in the country. He linked this epidemic in part to the lack of connection people have with each other, as well as with the climate crisis, and called on us to "cultivate a culture of connection." We helping professionals can explore why we often minimize or ignore the growing disconnection with our own bodies, each other, other life forms, and the ecosystems that sustain us. We can understand how this disconnection is linked to what McGilchrist (2009) called the combination of "impotence and moral superiority" that we can see in our offices and communities today.

In our work with the climate crisis, we can view human beings through what I like to call an *inclusive practice* perspective in which we address all the ecobiopsychosocialspiritual dimensions of being human;

grapple with climate at the micro, mezzo, macro, and eco levels of practice; and inform our work through the scientific, logical, intuitive, and imaginative ways of knowing. Over the past decades, the spiritual dimension, for example, has become increasingly integrated into the helping professions, and perhaps the best synonym for *spirituality* that has emerged in the literature is *connection*. Spirituality can be understood as an individual dimension of human development (as are the cognitive, social, emotional, and physical dimensions) in which we become increasingly conscious of our connection with everything in the universe (Derezotes, 2006). Helping professionals can thus understand consciousness and connection as fundamental parts of what Maslow (1971) called the "further reaches" of human evolution and thus as key developmental goals that can contribute to the survival and well-being of humanity.

A final connection we can explore is that between the many disciplines that can contribute to our understanding of human well-being in the climate crises. The relevant multidisciplinary literatures that address climate change are enormous, and each one is growing. No single text can offer the complex theories and discoveries being generated in the many relevant fields, including meteorology, physics, sociology, political science, economics, medicine, philosophy, psychology, geography, religious studies, evolutional biology, and geology. We can, however, identify the key connections between these literatures that are especially useful in informing our practice. For example, we can examine how the existing psychological literature on grief and loss can help inform our interventions with the grief most of us feel about the accelerating destruction of Earth's species and healthy habitats. Alternatively, we can examine how studies in economics and political science can help shine the light of truth on false concerns spread by powerful interest groups that we simply cannot afford to implement the climate solutions necessary today.

In spring 2023, I was kindly invited to write a text on climate change by Cognella Editorial Senior Vice President Kassie Graves. At the time, I was only weeks into "retirement" from academia and was still very much a recovering academic. I thought I had published my last text earlier that year. However, I felt an opportunity and recognized a responsibility. Like many of us, since I was a child, I have been concerned about the global survival threats we face today, including preparations for war, overpopulation, xenophobia, inequality, pandemic, and climate change. When I became an academic, I found ways to include these topics in my practice, teaching, and scholarship. Indeed, many of the ecobiopsychosocialspiritual symptoms that I noticed in myself and the people around me seemed to be associated with the deteriorating conditions we humans had largely created. I agreed to sign a contract to write the book.

The text is written for all helping professionals. A major theme in the text is that the climate change we want to see in the world begins with a climate change in human consciousness, which can then lead to the climate change we want in the "humansystems" and ecosystems we all live within and thus are a part of today. I built my work, of course, upon the work of many other people who came before me. I based the theory of *climate-sensitive practice* that is introduced in this text upon the theories of inclusive practice and advanced generalist practice that I developed in my previous textbooks.

Perhaps we can end the preface and begin the text with a story. There are apparently multiple versions of the "two wolves" story. I first heard this version from a Navajo man named Lacy whom I worked with years ago. A similar story can be found at LonerWolf (2023).

In this version, there was a beloved grandmother who was the wisest person in her village. One day she sat the children down and told them about the two wolves that live inside of her and every person. There is a beige wolf, who represents such traits as envy, resentment, sorrow, greed, ego, and lies. The other is the gray wolf, who represents such characteristics as joy, peace, love, humility, kindness, generosity, and faith.

The beige wolf, she told the children, was seen as evil and seemed to look for ways to cause harm to other living beings in the forest. The gray wolf was viewed as good and seemed to try every day to be of assistance and promote the highest good. Inside of each of us, the grandmother explained, the two wolves compete with each other to see which will be in charge of that person.

"Which wolf wins that fight?" asked the children.

The grandmother replied, "They both win because I feed both of them."

The children were confused, so the grandmother explained: "Although seen as evil, the beige wolf has useful contributions to make, such as will, tenacity, courage, fearlessness, and cleverness. The gray wolf can also contribute, with such characteristics as emotional strength and compassion and an awareness of the highest good. The two wolves need each other, and I need them both. If I fed only one of them, the other would become angry, bitter, and resentful and would only bring harm."

To me, the two wolves story is about inclusivity, consciousness, and connection. We need the cooperation and help of everyone to cocreate the best climate solutions. How we deal with the differences that divide our community will determine how well we can bridge the differences that now divide us and then cooperate to support the highest good for all. Just as we need to make friends with both wolves within ourselves, we also need to cooperate with and care for about all people, particularly those whom we fear or mistrust most and with whom we most disagree.

I believe that the wisdom of our collective whole, of all humanity, will resist the temptation to turn to the authoritarian strongmen who have been willing in every era to take advantage of turmoil, fear, and uncertainty by offering denial, hate, and violence as short-term solutions to the challenges humanity faces. These strongmen still exist today, of course, and are still willing to seek personal gain by any means necessary. I believe we will nevertheless collectively transform our families, villages, cities, institutions, and states into the kinds of inclusive and cooperative communities we need to survive and prosper.

We may sometimes stumble in our resolve to support the highest good, but the wisdom of kindness, justice, and inclusion will eventually triumph.

REFERENCES

Derezotes, D. S. (2006). *Spiritually oriented social work practice*. Pearson.

LonerWolf. (2023). *The two wolves story*. https://lonerwolf.com/two-wolves-story/

Maslow, A. H. (1971). *The further reaches of human nature*. Viking.

McGilchrist, I. (2009). *The master and his emissary: The divided brain and the making of the Western world*. Yale University Press.

U.S. Department of Health and Human Services. (2023). *New surgeon general advisory raises alarm about the devastating impact of the epidemic of loneliness and isolation in the United States*. May 3, 2023. https://www.hhs.gov/about/news/2023/05/03/new-surgeon-general-advisory-raises-alarm-about-devastating-impact-epidemic-loneliness-isolation-united-states.html

Wilson, E. O. (1984). *Biophilia*. Harvard University Press.

Introduction

... and no challenge—no challenge—poses a greater threat to future generations than climate change.

—BARACK OBAMA (WHITE HOUSE, 2015)

The climes, they are a changin'.

—MERRIAM-WEBSTER, N.D.

IMG 0.1

IMG 0.2

It is a Thursday morning in June 2023, and, sitting in front of my desktop computer, I am curious to see how the text will begin. The news is streaming in the background, as a dry, warm wind pushes into the room through the window screen. When I look outside, towering cumulus clouds are already rising over the mountains.

As I reflect, the screen displays hazy orange views of the unprecedented air pollution in New York City, choking with smoke blown south from hundreds of Canadian wildfires. These fires are the result of climate change–driven extreme heat and drought, explains the reporter, and she adds that our East Coast cities today have the "worst air in the world."

Then, the local weather segment arrives, and the meteorologist explains that we will have the same thunderstorms this afternoon.

"I'm sorry, but you know the drill by now," she says, smiling. "This spring we continue to have this unusual pattern of the monsoon-like weather that typically does not arrive in Utah until later in the summer. However, the good news is that the flooding from our record snowmelt in the Wasatch Mountains appears to be winding down."

Then, there is a commercial offering a new health product for sale, with a catchy name. We hear an enticing musical jingle, with images of people singing their promotional song while smiling and dancing and happily engaging in unusual activities. The screenshots shift quickly, apparently meant to further activate our sullen and lethargic attention spans. As the drumming and guitar riffs offer an upbeat musical background, a pleasant voice rapidly reads off a list of possible side effects of the product, which, if one actually listens carefully, include some potentially life-threatening physical reactions.

The next segment displays a tremendous surge of water draining through the gigantic breach in the Kakhovka Dam on Ukraine's Dnipro River. We are shown clips of people already traumatized by the Russian invasion and now trying to survive in the massive flooding downstream. Water supplies for farming and city water systems will be disrupted, we are told, and Ukraine has accused Russia of destroying the dam to block the expected Ukrainian summer counteroffensive. The reporter ends with a segment about how some early U.S. presidential candidates have taken a stand against refunding the Ukrainians, further fueling the political polarization in Washington. No mention is made of violent conflicts and climate-related flooding happening today in Africa and Asia.

I realize that everything displayed on the screen seems somehow related to the topic of our climate change crisis. Each news item is covering an aspect of climate in human consciousness, climate in our human systems, or climate change in our ecological systems. The news is mostly negative; we are told that conditions are generally getting worse. There is also an emphasis on immediate symptom reduction (relief of physical and psychological discomfort) rather than on the possibility of addressing the root causes of the immensities that now challenge us today. I did not know at the time that the following month, July 2023, would eventually be called the warmest month ever recorded on Earth, affecting at least 81% of the planet's human population.

Curious about the word *climate*, I look it up in the dictionaries and thesauruses. I learn that the meanings of the word have changed over time. The origin of *climate* apparently came from the ancient Greek word *klima,* which meant "inclination" or "latitude." Over time, people started to use the English word *climate* to describe weather systems. However, by the mid-seventeenth century, *climate* was also being used to describe the "atmosphere" of various human systems, such as those found in families, institutions, or nations. Today, *climate* is often paired with the word *change*, as in *climate change*. Finally, I learn that the word *climax* shares the same origins with *climate* and refers to "the point of highest dramatic tension or a major turning point in the action of a play, story, or other literary composition" (Merriam-Webster, n.d.). We certainly do seem to be approaching a climax crisis in both ecosystem climate change and the climate change in human consciousness and human systems as well.

So, the terms *climate* and *climate change* can be used to describe interrelated features in human activity and ecosystem activity. What are those features? In social work education, we often talk about the importance of values, skills, and knowledge. Perhaps we should start by looking at values that seem to be linked to climate change in both human activity and ecosystems.

Values

Grabbing a pencil and paper, I start to list the values that seem to inform the ways we relate to each other and with the ecosystems that sustain all life. That list became Table 0.1. The entries in column 1 describe some of our current, widely held values, which are associated with the ominous climate change crisis that we also call *global warming*. The phrases in column 2 represent the values that I believe would help us successfully address this crisis and cocreate a positive climate change in our human consciousness, humansystems, and ecosystems that supports our common good. In this context, *humansystems* include the formal and informal ways we humans interact with each other, in such arrangements as couples, families, institutions, neighborhoods, communities, and nations. *Ecosystems* include such natural structures as grasslands, taiga, deserts, tundra, rivers, lakes, littoral zones, coral reefs, temperate forests, and tropical rainforests. We put the word *natural* in quotation marks because we know that most ecosystems on Earth today have been at least moderately affected and altered by human activity.

TABLE 0.1 Climate Change Values

	Values commonly currently held in our culture that seem associated with global warming, pollution, and species extinction	Emerging cultural values that seem associated with deep peace, sustainability, global healing, and transformation
1	Passive focus on symptom relief	Assertive focus on activism
2	Sense of grandiosity and moral superiority	Sense of hubris, humility, healthy ego
3	Feeling of powerlessness, despair	Radical sense of hope and faith
4	Pervasive loneliness, community hunger	Discovery of connection with everything
5	Identification with wealth, power, fame	Identification with collective well-being
6	Tendency toward self-harm and suicide	Reconnection with body, love of life
7	Short-term individual comfort	Long-term global well-being
8	Aggressive "resentiment" toward the Other	Attitude of love for self and the Other
9	Denial of loss; depression and anxiety	Grief of loss, letting go, reinvestment
10	Emphasis on Western and majority voices	Inclusion of all voices
11	Focus on debate, making a point	Focus on dialogue, conformation of otherness

The 11 items in Table 0.1 are not listed in a particular order, and all are interrelated. Item 1 refers to a general passivity about our global survival today, associated instead with a focus on our physical and psychological symptoms and on a search for quick relief, rather than on what we can do to be of service to the world. We will be looking later in the text at how this passivity is linked not only with such widespread symptoms as anxiety and depression but also on with our propensities for self-harm, suicide, aggression toward others, and indifference to other living things and the ecosystems that support all life. The second item calls attention to our increasing left-brain and egoic dominance, associated with such symptoms as grandiosity, entitlement, and moral superiority, which has increased over at least the past two millennia (McGilchrist, 2009). We will explore how this sense of grandiosity, entitlement, and moral superiority is associated with the climate crisis and how these processes can be healed and transformed. In the text, we will explore the role of *humility* in this healing and transformative process, a word derived from the Latin *humus*, which means "the earth." To be *humble*, from this perspective, means to be rooted in the ground of our Earth, aware and accepting of our own strengths and limitations.

Item 3 refers to the unfortunate sense of powerlessness and despair that many of us have today. We tend to complain about the global threats to our existence and then shake our head and say, "I hope people get it together and do something, but I am not very optimistic." In this text, we will study the limitations and utility of hope, not only in our own individual well-being but also in our quest to address our local and global challenges. We will start to develop a theory of radical faith that includes both a positive attitude and activism. The loneliness and hunger for community that most people seem to experience today is the value in item 4. If I had a magic wand and could offer only one thing to the people I know, it would probably be an inclusive, diverse, and equitable community in which they could participate. We will look at how important a sense of connection with everything in the universe is to human well-being and how *connection* may also be the best synonym for a spirituality that many of us are seeking today. We will look at how healthy communities can contribute to the transformation of the climate crisis.

With the fifth item, we consider our identification with personal wealth, power, and fame. Identification is a primary function of the human ego. It is the egoic tendency to see something else as the same as me. For example, I might see a political party as not just something I support, but also something that actually *is* me, as I *am* also that. A big danger in identification is that I may react to any perceived criticism of what I identify with as an attack on me. When this happens, I can justify to myself the option of violence as a legitimate reaction. For example, if I have identified with my political party, I may attack you violently when I think that you criticized this party because it feels like you attacked me. We will explore the large role identification with wealth, power, and fame plays in the etiology of climate change and in human well-being. The reader will be asked to consider what it might take to help people move toward and identify with collective well-being, which includes the condition of all human beings, all other living things, and the ecosystems that support life.

We will also explore our tendency to do self-harm and to engage in suicidal thoughts and actions, which is the sixth item. We are aware of the rising rates of self-harm and suicide in our populations and of how common self-harm is in the everyday life of most people. When I ask my students for examples of everyday self-harm, they can easily name any number of self-destructive activities they see in themselves and others. For instance, many of us eat poorly, consuming highly processed foods and beverages that are dangerous to our health. Many of us choose to not sleep for enough hours, despite the knowledge that this, too, is a dangerous activity, confirmed by the chronic exhaustion we feel every day. When I ask my students how they view their own bodies, they often admit that they tend to see their bodies as machines and forget that they are not just brains but also animals with bodies that require proper care and have things to teach us. In the text, we will explore how our relationship with our own bodies, including our brains, is often rather disconnected and confused. We seem to often expect our bodies to perform perfectly, without a need for rest, repair, and care, and we may become angry when our bodies have symptoms such as illness or pain. We often feel a disconnection from our brain but also report an inability to deal with the endless stream of anxiety that seems to come from it. We will look at how our relationship with our bodies (the bio-psycho-social) is associated with our relationship with other living things and our ecosystems (the eco-bio-psycho-social).

The seventh item addresses the interrelationships between our human desire for short-term comfort, human well-being, and our climate change crisis. Our tendency to seek pleasure and avoid pain seems to be part of our nature, as noted in many world wisdom traditions. It can feel very difficult to face the challenge of climate change, and the remedies, as we will see, require long-term changes in the way we think and act. It may seem far easier to complain about how bad the weather has become and to retreat into our distractions. In the text, we will explore how we might heal and transform these human tendencies for our collective good. We will also explore the theory of *resentiment,* which is the eighth item on our list. Why are so many of us across the world full of anger and aggression and willing to blame the Other for our own suffering? We will explore this important topic, how this ongoing anger or resentiment is associated with our climate change challenges, and how we professional helpers can work to help heal and transform resentiment.

We know that when a human being experiences loss, they will need to go through a grieving process to heal that usually includes *feeling* our loss. However, in our culture, sadness is often equated wrongly with depression and often denied or minimized. In item 9, we take a look at the inability or unwillingness of the world to grieve the loss of clean air, clean water, wild lands, and diversity of wildlife and the appearance of other symptoms of the damage we have done to other living things and our ecosystems. By helping people to feel and heal the loss, we may also help them to move to a place where they can take action to help reverse the climate change that we have already cocreated.

There is a tendency in the United States to give the strongest voice to populations in the Western world and minimize or ignore the voices of other populations. In other words, as item 10 suggests, there are significant climate justice issues in local and global communities. We will examine how the people most harmed by the climate crisis are often members of populations that are already minoritized, vulnerable, and voiceless.

Finally, in the last item, we look at an important communication tool we have to deal with ecosystem climate change, which we call *dialogue.* Many of us attended schools where there was a debate team, in which people compete to win a conversation. Very few of us attended programs that offered dialogue classes or teams. As we will study, the purpose of dialogue is to create relationships that can bridge the many differences that divide us today. We will examine ways to use dialogue skills to help people develop the cooperation necessary to successfully transform our climate crisis.

The Climate Crisis

What is the climate crisis? According to the United Nations (UN), we now have three interrelated global environmental crises: *climate change, biodiversity loss,* and *pollution and waste* (UN News, 2021). Since these crises share common etiologies and outcomes, in this text, we will use the phrases *climate crisis* or *ecosystem climate change* to include all three crises. We will use the phrases *healing climate change* or *transforming climate change* to represent our efforts to do healing and transformational work in our own consciousness, humansystems, and ecosystems.

What general content should go into a textbook about ecoclimate change for social work and such related fields as psychology, nursing, psychiatry, and educational psychology?

The text should assist professionals in these fields to address the climate change in ecosystems crisis in their practice. We want to learn how to:

1. Assess the impact of the crisis on our clients
2. Engage with populations in dialogue about the crisis to foster cooperative relationships

3. Intervene to help our clients not only cope effectively with the crisis but also become effective climate activists who can work effectively with other people
4. Assess the outcomes of our work and make changes when necessary in our helping strategies

The Text

What Is the Purpose of the Text?

First, we want to help foster and sustain a transformational climate change in human consciousness, which we can see has already begun. *Consciousness* can be defined as reverent awareness, in which we increasingly see things the way they actually are, with more friendly (accepting) eyes. Seeing the damage and suffering that exists today in our own consciousness, humansystems, and ecosystems with friendly eyes does not mean we have to like what we see, but rather that we accept that in this moment, this is what we are dealing with. Therefore, seeing things with acceptance allows us to grieve for what we have lost and eventually move to a place where we can become committed activists and be of service to our world.

We want to face together the realities of global warming, pollution, and loss of biodiversity. We want to experience firsthand the connections between ourselves, other living things, and the ecosystems that support all life. Consciousness practice is truth-seeing, truth-telling, and, as we will examine, the foundation of how we can help promote transformational climate change in our humansystems and ecosystems. Such work should be guided by our commitment to social justice, in which the principles of equity, diversity, and inclusion inform all our actions. This goal is represented by the box labeled "Climate Change in Consciousness" in Figure 0.1.

Second, we want to help foster a transformational climate change in humansystems. This work addresses the eco-bio-psycho-social-spiritual climate of the individuals, families, institutions, and local and global communities of our species. The climate crisis that threatens us today is associated with the way we human beings interrelate with each other. Our efforts to transform our humansystems should be guided by our commitment to social justice in which the principles of equity, diversity, and inclusion inform all of our actions. In order to make these transformations, we also need to find ways to cooperate across the differences that currently divide us. Research shows that most people in the United States and, to varying degrees, most other countries are aware of and concerned about the climate crisis. We also do not yet always agree, of course, on the best ways to respond. Across the world, there tends to be a polarization between people with different views on climate in both our local and global communities. This second purpose is represented by the box labeled "Climate Change in Humansystems" in Figure 0.1.

Third, we ultimately want to also cooperate together to foster a transformational climate change in the ecosystems that support all life on Earth. We can do so by modeling and otherwise promoting practices that can contribute to the well-being of human life, all other living things, and ecosystems that support all life. A climate change in our ecosystems builds on the climate change in human consciousness and in human systems that we examine in sections 1 and 2. This third purpose is represented by the box labeled "Climate Change in Ecosystems" in Figure 0.1.

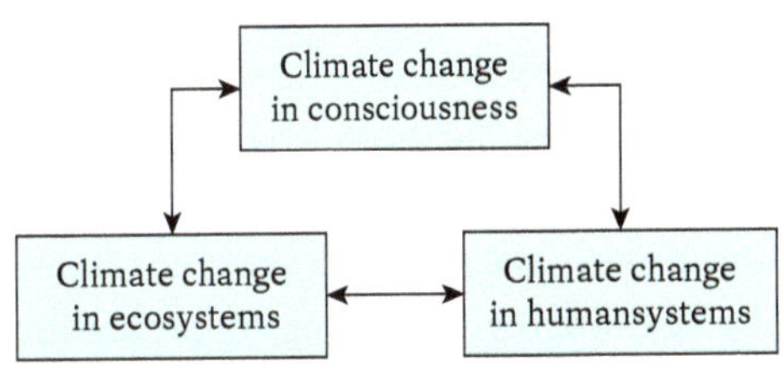

FIGURE 0.1 Three Interrelated Components of Transformational Climate Change Practice

This book is written as a practice text for social work and related fields. Since most readers will be psychotherapists, counselors, case managers, teachers, nurses, health practitioners, facilitators, and other professional helpers, there is an emphasis upon practical approaches to transformation of human consciousness and transformational climate change in humansystems and ecosystems. Although the most relevant scientific facts will be summarized, we will not examine the detailed contributions from the many scientific disciplines now working hard to help us understand what is going on with our other living things, pollution, and global weather. Therefore, in the text, I will summarize the essential scientific background when needed, and since the relevant literature is already vast and quickly growing, the technical descriptions in the text will be as concise as possible, without oversimplification. We will infuse references to climate change throughout Section 1, and these entries will be presented in "Science Resources" boxes, the first two of which you will see later in this introduction. Students are encouraged to continue to seek further knowledge from the literature on their own.

The three sections of the text will reflect the three interrelated components of transformational climate change practice illustrated in Figure 0.1.

In **Section 1**, readers are challenged to work with their own *consciousness change*, which we define as seeing themselves and their world more accurately and with greater acceptance. We want to teach students and practitioners a consciousness practice in which we learn to see our world and ourselves more accurately. We also want to raise their consciousness about how ecoclimate change is a complex combination of global warming, pollution of water, earth, and air; and biodiversity loss that has accelerated in response to human activity. We also want to help students and practitioners raise their consciousness about how their individual well-being is associated with that of other humans in their local and global human community, other living things, and the ecosystems that support all life. We will utilize an inclusive approach to practice that uses interventions in all seven paradigms of work.

In **Section 2**, we address transformational climate change at the humansystem levels of family, institution, local community, and global community. We introduce the concept of climate-sensitive practice that can be used by all the helping professions when working at these levels. Human system climate change utilizes both conscious work and dialogue practice. We also study how we can address such climate justice concerns as inequity of inequity of power and resources as well as the extreme vulnerability of many minoritized populations to climate crisis-related harm. The causes and possible solutions to the political polarization that currently often inhibits our progress are also explored.

In **Section 3**, readers learn about practice approaches that can help transform our global and local biomes and anthromes. Whereas an ecosystem includes the associations between living and nonliving things in an environment, a biome is a specific geographic area classified according to the living species that exist there and can consist of many ecosystems. Some scientists classify biomes broadly, using such terms as *grassland*, *forest*, *desert*, *tundra*, *freshwater*, and *marine*; other scientists use more specific and local classifications. *Anthromes*, in contrast, are areas that have been influenced by ongoing interactions between people and ecosystems and may include urban, village, cropland, rangeland, and cultured anthromes (National Geographic, 2023a). In this last section, we offer interventions that can promote the sustainable well-being of human life, other living things, and our ecosystems, in all our planetary biomes and anthromes.

TABLE 0.2 Plan of Book

Sections	Chapters
Section 1: Climate Change in Consciousness	Chapter 1: Inclusive Practice With Human Consciousness Chapter 2: Reconnecting With Our Inner Nature: Healing Body, Mind, Spirit Chapter 3: Healing Our Ego-Hemispheric Imbalances Chapter 4: To Be or Not to Be: Eros and Thanatos Chapter 5: Despair, Hope, and Faith-Based Activism
Section 2: Climate Change in Humansystems	Chapter 6: Inclusive Practice With Humansystems Chapter 7: Polarization and the Radical Middle Chapter 8: Climate Justice: Practice in Diversity, Equity, and Inclusion Chapter 9: Climate-Sensitive Practice With Families and Institutions Chapter 10: Climate-Sensitive Practice With Local and Global Communities
Section 3: Climate Change in Ecosystems	Chapter 11: Inclusive Practice With Ecosystems Chapter 12: Earth, Wind, Fire, and Water Chapter 13: Practice With Biomes Chapter 14: Practice With Anthromes Chapter 15: Conscious Climate Change

When I have facilitated intergroup dialogues about the climate change crisis, there are a few questions that usually come up. The first is, is there really a climate crisis? The short answer is yes. Climate change is different from weather change. The term *weather* is used by scientists to describe short-term conditions we experience each day, whereas the term *climate* is used to describe long-term conditions. See Box 0.1 for more information.

SCIENCE RESOURCES 0.1 Is there really a problem? Is ecological climate change real?

Climate change is an empirical fact, well described in the literature; scientists know that the average temperature of Earth has been rising quickly since the beginnings of the industrial era and is still accelerating today (Krauss, 2021).

A recent analysis of climate justice and safety by the Earth Commission (Rockström et al., 2023) found that the well-being of our planet is "failing in seven out of eight key measures." The authors proposed (p. 1):

> safe and just Earth system boundaries (ESBs) for climate, the biosphere, fresh water, nutrients and air pollution at global and subglobal scales . . . that span the following major components

of the Earth system (atmosphere, hydrosphere, geosphere, biosphere and cryosphere) and their interlinked processes (carbon, water and nutrient cycles).

We know that the diversity of our animal and plant life on Earth is rapidly decreasing; in fact, so many species have died off that some scientists have even proposed that we are living during one of the great extinctions that have occurred on the planet. This current sixth extinction is sometimes called the *Holocene extinction*, named after the current geologic epoch we live in today, which began about 11,700 years ago after the last major ice age declined. Some scientists use the term *Anthropocene extinction* (using the Greek word *anthropo*, for "man"), which could be thought to have begun during the Industrial Revolution of the 1800s, when human activity increased carbon and methane levels in Earth's atmosphere, or in 1945, when the first nuclear bombs were detonated. Our atmosphere, oceans, freshwater, and landscapes, as well as our own living spaces and bodies, are becoming more polluted with substances that threaten the well-being of all life on Earth (National Geographic, 2023a; 2023b).

The Intergovernmental Panel on Climate Change (IPCC) is the section of the UN assigned to assess the science related to climate change. Its recent report (IPCC, 2023) summarizes some of the most profound scientific evidence substantiating the climate crisis. The report states that there is now more carbon dioxide in the atmosphere than at any other time in our history. Human activity has already raised average global temperatures by 1.1 degrees and is rapidly moving us toward the threshold scientists set, at 1.5 degrees Celsius of warming. About two-thirds of extreme weather events in the past two decades are associated with human activity, and these events continue to be more frequent and extreme. Additionally, our global wildlife populations have dropped by 60% in just over 40 years.

People also often want to know if it is true that humans have caused this our climate crisis. See Box 0.2 for more information.

SCIENCE RESOURCES 0.2 How much of the climate crisis is human-caused?

IMG 0.4

Our global climatic conditions have changed dramatically since levels of human-generated carbon production accelerated during the Industrial Revolution. Scientists have discovered this association by studying a wide range of complex factors, including the overlapping planetary cycles of Earth, the reflectivity of the planet's surface and atmosphere, the chemistry of the so-called greenhouse effect, details on ancient temperatures, and atmospheric chemistry. Such data have been entered into increasingly detailed and accurate supercomputer climate models (Krauss, 2021).

Although we know that Earth's climate has been in constant change over millions of years, there is strong evidence that the Earth's average temperature has warmed due to human activity. The burning of fossil fuels is a primary cause of global warming, and human activity is also the primary causes of related pollution and species extinction (UN, 2023).

Climate science demonstrates that human activity is the major cause of global warming, using data from such disciplines as oceanography, meteorology, chemistry, physics, biology, and computer science. They compare observed and modeled patterns of weather to identify the "human fingerprints" that indicate human-caused contributions to warming. These fingerprints have been found in data collected in our atmosphere, oceans, and Earth's surface and include carbon dioxide increases, historical warming, and atomic signatures in the atmosphere from fossil fuel combustion (Union of Concerned Scientists, 2021).

Finally, I write this book with a faith-based activism that we can and will do the healing and transformational work necessary to turn things around. As we will explore further in Chapter 6, *faith-based activism* can be defined as a combination of radical hope and committed involvement. Climate activism will no doubt often continue to be difficult work, and we will make mistakes. We may continue to fight among ourselves about the causes, nature, and solutions of the climate crisis. Perhaps we will see things get worse on Earth in some ways before they get better, and we may sadly see many additional people and other living beings suffer even more from the climate crisis before we make substantial progress in transformational climate change. But human beings have rallied before after other challenging events in our past. Life was never easy for our ancestors; it can be hard to be a human.

However, it can also be wonderful being a human being. We get opportunities every day to have the pleasure of living in this beautiful universe and to make a positive difference in the world.

QUESTIONS FOR REFLECTION

1. The values each of us hold can be placed in a hierarchy in which the top value is the one most highly held, the second on the list is the second most highly held, and so on. Take a look at column 1 in Table 0.1, emerging cultural values that seem associated with deep peace, sustainability, global healing and transformation. Then, put them in the order that you most personally rate them. There is no "right" answer to this exercise, of course; it is about your personal preferences.

2. Please take a look at Figure 0.1. The text suggests that these components are connected. Do these three kinds of climate change, illustrated in the figure, actually seem to be interrelated to you? Please explain.

3. The idea of faith-based climate activism that combines hopeful attitude and skillful activism is introduced in this introduction, to be explained later in the text. What is your attitude toward our climate crisis? Do you believe that humanity will deal with the crisis effectively in your lifetime? Do you believe that human beings will still survive and be living on Earth in 500 years? Please explain.

REFERENCES

Derezotes, D. S. (2023). *Inclusive social work: A new vision of community practice.* Cognella.

Intergovernmental Panel on Climate Change. (2023). *AR6 synthesis report: Headline statements.* https://www.ipcc.ch/report/ar6/syr/resources/spm-headline-statements/

Krauss, L. M. (2021). *The physics of climate change.* Port Hill Press.

McGilchrist, I. (2009). *The master and his emissary: The divided brain and the making of the Western world.* Yale University Press.

Merriam-Webster. (n.d.). *"Climate" change.* Wordplay. https://www.merriam-webster.com/words-at-play/climate-word-origin

National Geographic. (2023a). *Education resource: Biomes.* https://education.nationalgeographic.org/resource/biomes/

National Geographic. (2023b). *Education resource: Anthropocene.* https://education.nationalgeographic.org/resource/anthropocene/

Rockström, J., Gupta, J., Qin, D., Lade, S. J., Abrams, J. F., Andersen, L. S., Armstrong McKay, D. I., Bai, X., Bala, G., Bunn, S. E., Ciobanu, D., DeClerck, F., Ebi, K., Gifford, L., Gordon, C., Hasan, S., Kanie, N., Lenton, T. M., Loriani, S., ... Zhang, X. (2023). Safe and just Earth system boundaries. *Nature, 619*, 102–111. https://www.nature.com/articles/s41586-023-06083-8

UN News. (2021). *SDGs will address "three planetary crisis" harming life on earth.* United Nations, April 27, 2021. http://news.un.org/en/story/2021/04/1090762

Union of Concerned Scientists. (2021). *How do we know that humans are the major cause of global warming?* https://www.ucsusa.org/resources/are-humans-major-cause-global-warming

United Nations. (2023). *Climate action site.* https://www.un.org/en/climatechange/reports

The White House. (2015). *Remarks by the president in state of the union address.* January 20, 2015. https://obamawhitehouse.archives.gov/the-press-office/2015/01/20/remarks-president-state-union-address-january-20-2015

Credits

SECTION 1

Climate Change in Consciousness

IMG 1.1

IMG 1.2

The main thesis of this text is that the roots of our climate crisis can be found in our own human consciousness, and as we transform our consciousness, we can help heal and transform the climate of our interrelated humansystems and ecosystems.

What is consciousness? Essentially, human consciousness is reverent awareness of myself and the world in which I live. Human awareness is subjective, meaning that my experience of myself and the world is influenced by my own psychology, with all its conditioning, past experiences, beliefs, and emotions. Reverent awareness occurs when I see things with friendly eyes, accepting what is in this moment, which does not necessarily mean that I like or dislike what I see.

A climate change in consciousness occurs when I become aware of how my mind influences my consciousness. With such awareness, I see myself and the world more accurately. We will look at how when we heal, we become whole, and how the healing of consciousness involves this expansion of our reverent awareness. In the last chapter of the text, we will also explore the relationship of human consciousness with a universal consciousness that our ancestors often sensed and that is now being theorized by both scientists and philosophers today.

Our long-term well-being and ultimate survival as a species is associated with our willingness to change the way we view and interact with ourselves, each other, other living things, and the ecosystems that support all life. As we will further examine, when we make a transformation, we change the forms of our emotions, thoughts, actions, or products we create. The transformation of consciousness involves developing the ability to see the world without interference from the beliefs and thoughts we have from the past so that we can create new attitudes and ways of being in the world that support the highest good.

In Section 1, which includes Chapters 1 through 5, we will look in depth at the key elements of **human consciousness** that are associated with climate change in humansystems and ecosystems. In Chapter 1, we will examine how we can engage, assess, and intervene with our consciousness from an inclusive practice perspective. In Chapter 2, we address the fundamental concept of *connection*, which can also be understood as a synonym for our spirituality. Our loss of awareness of the fundamental connection we have with everything in the universe is associated with the ecosystem climate change crisis we face today.

In Chapter 3, we will examine the functions of ego, hemispheric brain dominance, and grandiosity and various ways to work with these functions in our practice. Chapter 4 explores the life and death drives in humans, using the archetypal psychology of Eros and Thanatos. In Chapter 5, we study the topics of hope and despair and offer a faith-based activism approach to human consciousness.

Credits

IMG 1.1

Inclusive Practice With Human Consciousness

Hey, mercy, mercy me …
What about this overcrowded land?
How much more abuse from man can she stand?

—MARVIN GAYE (1971)

Climate change refers to long-term shifts in temperatures and weather patterns. Such shifts can be natural, due to changes in the sun's activity or large volcanic eruptions. But since the 1800s, human activities have been the main driver of climate change, primarily due to the burning of fossil fuels like coal, oil and gas. … Energy, industry, transport, buildings, agriculture and land use are among the main sectors causing greenhouse gases. … Climate scientists have showed that humans are responsible for virtually all global heating over the last 200 years. … The consequences of climate change now include, among others, intense droughts, water scarcity, severe fires, rising sea levels, flooding, melting polar ice, catastrophic storms and declining biodiversity.

—UNITED NATIONS (N.D.)

I was born into a Greek-Italian family on the South Side of Chicago. I still remember the big porches facing the streets, where I spent a lot of time, and a big house full of people. When I was 6 years old, my parents separated physically and psychologically from the extended family and moved us to a suburb on the northwest side.

Our new house was a wooden rectangle built on the "rolling meadows" of newly planted and thirsty Kentucky bluegrass, one of a number of enclosures planted along a newly laid street that seemed to a little boy to go on forever. One of my earliest memories is of sitting on the front step of our new home and crying.

The heart often knows what is going on before reality registers in the brain. What were those tears of mine about? What had been lost?

Lost were the friendly, welcoming porches, clustered together, which had been replaced by garage doors that were neither inviting nor connecting. Gone was the familiarity of our own culture and extended family and of neighbors whom we had known forever. We had lost community.

Gone also was the wealth of diversity in our home city. We had become part of what sociologists called the *White Flight* of the 1950s—an exodus of fear, seeking escape as more people of color, especially African American families, moved to Chicago, usually from the Deep South, seeking more freedoms and new opportunities. And, as has been true throughout history, when actions arise primarily out of fear, we tend to experience unwanted outcomes. In our flight, we helped create an even more segregated cityscape that had unforeseen ecobiopsychosocialspiritual consequences.

We lost the Northern Illinois meadows, streams, and lakes, which had suddenly become landscaped freeways, malls, and homes. We were now part of a vast experiment, initiated by privileged classes, who sought to buy and own a part of nature, only to find that we had "paved paradise and put up a parking lot" (Mitchell, 1970). We did not see yet that one cannot buy a part of nature without having an impact upon Her and thus undermining the very thing you wanted to own and protect.

We also lost equity as the wealthier classes were able to live in parts of town that were relatively less polluted and crowded. And the loss of equity tends to bring increased fear in privileged classes of those other people who have less. We were living in what was called "a good neighborhood" with "good schools"; yet, in our search for security and safety, we experienced a growing sense of anxiety, insecurity, and danger that remains today.

What Is Inclusive Practice?

The theory of inclusive practice (Derezotes, 2023) is introduced throughout this text. This theory was developed to assist practitioners in working with people at the micro, mezzo, macro, and eco levels of practice and to help address the connections between the well-being of people, of other living things, and of ecosystems on the planet. The climate crisis is one of the interrelated global survival threats (GSTs) that inclusive practice addresses, which also include preparations for mass-casualty war and terror, pandemics, and human overpopulation and xenophobia. In Sections 2 and 3 in our text, we will introduce the theory of climate-sensitive practice, which applies inclusive practice principles to the climate crisis.

The most advanced theory is the most inclusive, as my colleague Au-Deane Cowley would often say. Inclusive practice was developed to provide practitioners with a theoretical framework that supports professional helpers as we address the interconnected ecobiopsychosocialspiritual challenges that we all face today. Traditional approaches to what we still call *mental health* have especially emphasized biological etiologies, such as genetics and brain chemistry, as well as the intrapsychic processes that make up the descriptions of the vast majority of what we call *diagnoses of mental disorders*. An inclusive practice framework expands the idea of mental disorders into the more comprehensive idea of eco-biopsychosocialspiritual well-being. The etiologies of well-being are seen as including interpsychic processes in our families, institutions, and local and global communities as well as our relationships with other living things and ecosystems. We now have increasing scientific evidence that individual human well-being is, in fact, strongly associated with the well-being of all living things and our shared living environments.

Inclusive practice addresses all the ecobiopsychosocialspiritual dimensions of human development, which are summarized in Table 1.1. As you can see, in this schema, we emphasize the *relationship* we have with each of the five dimensions. The practitioner is encouraged to assess in all the developmental areas in every case and to determine which developmental areas are strengths for the client. For example, a particular client may have relatively high emotional and social intelligence but relatively moderate cognitive intelligence.

TABLE 1.1 Ecobiopsychosocialspiritual Assessment

Dimension	Basic Focus	Examples of Assessment Questions
Eco (ecological)	Our relationship with our human-made and "natural" environments	What is my attitude toward the climates of my living and work cultures? How do I react to the quality of my local air, water, and soil?
Bio (biological)	Our relationship with our body	How do I feel about my body? Do I listen to my body? If so, what do I hear?
Psycho (psychological)	Our relationship with our emotions and thoughts	Am I aware of my emotions and thoughts? How do my emotions and thoughts influence me?
Social	Our relationship with other people	What is my attitude toward other people? How do I relate to people outside the group with whom I Identify?
Spiritual	Our relationship with our own spirituality	Do I experience a connection with other people, other living things, and ecosystems? Does my life have a purpose?

Inclusive practice uses all human ways of knowing in assessing and evaluating practice. These include science, logic, intuition, and imagination (McGilchrist, 2009; see Table 1.2). The spirit of this schema is to encourage practitioners to use *multiple* ways of knowing in each case as a means of checking conclusions that are derived from any one of the ways of knowing. For example, a practitioner might first have an intuitive reaction to a client they meet, such as fear or anger. Then, the practitioner might consult the literature, using *evidence-based practice* (EBP) to determine whether there are studies that would give them additional insight into the population and issue they are addressing with the client. They might also ask the client questions that would help clarify what is going on, which is a technique in *practice-based evidence* (PBE). When the same conclusions can be collaborated by multiple ways of knowing, the practitioner can be more confident in their assessment.

TABLE 1.2 Four Ways of Knowing

Ways of Knowing	Basic Focus	Examples of Questions That Might Be Used for Each Way of Knowing
Science	The use of testable descriptions and predictions to build and categorize knowledge	What does the latest meta-analysis of the research on this population and issue tell us? How can I use both PBE and EBP techniques to assess?
Reason	The use of logical inferences to understand the world	How can we move from the premises and data we have now to some logical conclusions about this population and issue?
Intuition	The use of "all-body" knowing without dominant cognitive processing	What sensations and emotions do I notice in my body when I work with this client or interact with other living things?
Imagination	The use of intentional creativity to see what "could be"	What kinds of potential future options can I envision for the client in the short and long term? What kind of cultural climate changes do I want to see in the world?

The processes of EBP and PBE are contrasted and explored by Duncan et al. (2010) in their text *The Heart and Soul of Change: Delivering What Works in Therapy.* Whereas EBP draws upon disseminated scientific evidence typically about groupings of populations and methods, PBE uses evidence collected during interactions with our own clients.

Inclusive practice also uses all of the paradigms of intervention. The evolution and characteristics of these paradigms are described in *Advanced Generalist Social Work Practice* (Derezotes, 2000). As summarized in Table 1.3, each paradigm is focused upon one or two of the dimensions of development we explored in Table 1.1.

A key idea in this schema is that a practitioner can use interventions drawn from any or even all of the paradigms in the same case. The practice paradigms are not mutually exclusive, and none is judged to be overall superior or inferior to the others. Practitioners use an ecobiopsychosocialspiritual assessment to help them determine which interventions to choose that best fit the client with whom we are working. Generally, a strengths-based approach is used, in which the practitioner begins by identifying and building on the client's developmental *strengths*. Thus, for example, if a client presents with developmental strengths in the areas of cognitive and social development, the practitioner might begin by using some cognitive-behavioral and psychodynamic interventions, perhaps in both individual and group (i.e., social) sessions. Eventually, as progress is made, other interventions that are drawn from other paradigms may be tried as well.

TABLE 1.3 Paradigms of Social Work Practice

Paradigm	Basic Goal	Developmental Dimensions Most Addressed	Examples
Psychodynamic	To address past trauma so that I can better meet my needs now	Cognitive and social	Object relations, self-psychology
Cognitive-behavioral	To change the way I think and act in the here and now	Cognitive and social	Acceptance and commitment therapy, dialectical behavior therapy
Experiential-phenomenological	To increase awareness and effective expression of emotions	Emotional	Gestalt, person-centered
Transpersonal	To experience my connection with myself, other people, other living things, and ecosystems	Spiritual	Mindfulness, yoga, prayer
Case management	To receive advocacy, Information, and referrals	All	Help with housing
Biopsychosocial	To gain awareness, acceptance, and healing of physical body	Biological	Yoga, aerobic exercise
Ecobiopsychosocial (or deep ecology)	To take response-ability for the well-being of other people, other living things, and ecosystems	Ecological	Community gardening, nature activism

Source: Derezotes, 2023.

Ecosystem climate change is just one of a number of interrelated GSTs that challenge us today. The inclusive practice approach emphasizes the interrelationship between all these GSTs and the well-being of people, other living things, and ecosystems on our planet. Table 1.4 illustrates the current four major GSTs. Many of the assessment and intervention strategies used in this text can be used in dealing with all four GSTs.

These GSTs are interrelated because the current climate of our human consciousness is associated with the roots of all of them. For example, as we will explore, the human ego, when largely unconscious and therefore dominant in a person or community, contributes to pandemics, wars, the climate crisis, and xenophobia. Additionally, the GSTs are also interrelated in as much as they tend to reinforce each other. Thus, for example, our xenophobic fear of strangers and other people who seem different makes the cooperation we need to deal effectively with pandemics, wars, and the climate crisis much more difficult.

TABLE 1.4 Major Global Survival Threats

Global Survival Threats	Brief Description	Examples
Pandemic	Diseases that can be transmitted and kill many human beings	Coronavirus disease 2019 (COVID-19) Other diseases that could become pandemic include influenza, plague, Ebola
War	Use of mass casualty weapons for political means	Nuclear war Space-based weapons War using robots guided by artificial intelligence
Climate crisis	Largely human-caused global warming, ecosystem degradation, and species extinctions	Rising ocean levels Extinctions of coral reef ecosystems More extreme droughts, fires, storms, flooding Urban smog
Xenophobia and inequality	Social justice issues relating to equity, inclusion, and diversity	Growing inequality of wealth, power, and resources Political polarization related to culture wars Growing refugee populations

Inclusive practice also avoids dualistic approaches to assessment and intervention, and instead seeks to find a *radical middle* that includes valid insights and approaches drawn from both extreme positions on a continuum. In Figure 1.1, there are two examples of a radical middle position that can be found on a continuum with two extreme positions, one at each end. In the first example, we could propose that a radical middle position between the statements "All Democrats are evil" and "All Republicans are evil" would be "There is good and bad in everyone." Similarly, in the second example, if we look at the current debate about the existence and cause of climate change, we can name two extreme positions. The extreme position, "There is no human-caused climate change," is not supported by the evidence. However, the other position, "All ecosystem climate change is completely human-caused," cannot be supported either because some "natural" factors in climate change (e.g., volcanic events, the El Niño–Southern Oscillation, and cycles in the Earth's orbit and tilt) may have an influence on current global warming. Therefore, we could suggest a radical middle position, which is supported by scientific evidence: "There is a current climate crisis that is mostly human-caused."

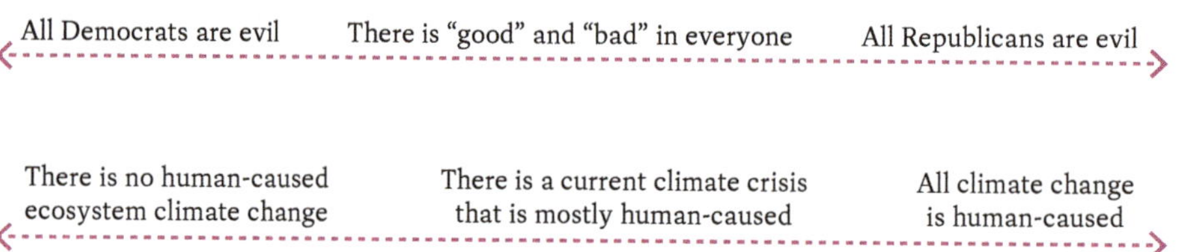

FIGURE 1.1 Two Examples of a Radical Middle Position Between Two Extreme Positions

Other Principles in Inclusive Practice

Healing and Fixing

Most experienced practitioners in the helping professionals know that they seldom, if ever, can actually fix their clients. What we mean by this is that we learn not to view the client as broken and helpless and a problem to be solved, nor do we expect that we can make their symptoms permanently go away. Rather, we view our work more as healing work. The English word *heal* literally means "to make whole" and was derived from the prehistoric Germanic word *khailiz*. We can understand our "wholing" work today by seeing professional helping as the process of helping people become more conscious of who they are, see their conditioning from the past, and learn to love and express themselves in ways that support the highest good. Effective practitioners know that all human beings experience at least some conditioning from their families, cultures, and peer groups and that a portion of this conditioning can be harmful to our health. For example, girls may still be taught today to never show anger, and boys may still be told not to cry. As people become more whole, they become conscious of their "parts" and learn to identify which come from conditioning and which are harmful. They also gradually learn to express themselves in ways that benefit the highest good, which includes the well-being of themselves, other people, other living things, and our ecosystems.

Applied to our work with the climate crisis, we see we cannot immediately or permanently fix the world for our clients, nor do we need to try to fix our clients, because they are not broken. When we strive to support our clients' healing, we help them further develop their consciousness so that they become more aware and accepting of who they are and of the nature of the world with friendly, accepting eyes. Although we, of course, strive to reduce unnecessary suffering, we also view suffering as a condition that can also get our attention and help make us more conscious. For example, as young people become more conscious, they become more aware and accepting of the many emotions that they may have in relationship with the climate crisis, including sadness, fear, and anger.

This distinction between fixing and healing can have profound implications for the goals of helping professions. In our introduction, we introduced some values that might help us address the climate crisis. Our values, placed in a hierarchy or an order of importance, give us our general direction, while our goals provide specific things we want to accomplish or attain. For psychotherapists and counselors living in the climate crisis era, for example, our primary goal shifts away from symptom reduction and elimination toward the development of consciousness and service in the highest good. Psychotherapists still try to help clients manage their suffering but also recognize, as we will see, that suffering can be the best teacher of consciousness.

For educators, we move away from the primary goal of transmitting knowledge and instead emphasize more the development and the use of consciousness for the highest good of individuals and communities. Students still learn knowledge, values, and skills, but the educator recognizes that without consciousness work, such goals are not sufficient to help students successfully address their lives during our climate crisis. We ask students to let their consciousness help inform the major decisions society needs to make today. If we ask students, for example, how our health care system might look if the well-being of all life and ecosystems were the most urgent goal, they might determine that the elimination of toxins in our environments must become a much higher value and goal than it is today.

Similarly, professional helpers may also move away from a primary goal of "fixing" individual physical symptoms and more toward an awareness of what human well-being includes and how it is interconnected with the well-being of all living things and the world's ecosystems that keep us all alive. In other words, physical health and mental health may best be understood as encompassing all the dimensions of ecobiopsychosocialspiritual well-being.

Transformation

We will also use the word *transformation* in the text when describing inclusive interventions. The word *transformation* is derived from the "Church Latin" word *transformationem*, which means "change of shape." Thus, when we strive to help transform our human-systems, our goal is to change the form of that system into a new form that would better support the highest good. For example, we may oppose politicians who seek to limit or destroy any governmental regulations over oil and gas production and instead help support efforts to reduce emissions from automobiles, planes, ships, and power stations that further pollute our soils, streams, and oceans. A transformation of our political landscape might include both reasonable policies that protect humans and other living things from dangerous emissions as well as perhaps an agreement to avoid the politicization of reasonable policies that protect ecosystems' well-being.

Micro-, Mezzo-, and Macro-Level Practice and Beyond

In an inclusive practice approach, the work we do may involve assessments and interventions at all four levels of practice: micro, mezzo, macro, and eco (see Figure 1.2). In conventional practice theory, micro-level work uses direct intervention with individuals, families, and small groups, often to assist with personal relationships and intrapsychic conflicts. In contrast, mezzo-level interventions target larger groups, organizations, and communities, often for the purpose of strengthening relationships between individuals and their communities. The focus of macro-level work is on systemic issues affecting local, state, national, and global communities.

We can add a fourth dimension of practice that is concerned with other living things and ecosystems. The words *micro* and *macro* have Greek origins. The word-forming element *eco* was invented by German zoologist Ernst Haeckel as *Ökologie*, which also comes from a Greek word, *oikos*, which means "house, dwelling place, habitation." We use the prefix *eco* to describe work that addresses the well-being of all living things and ecosystems, in relationship with human beings.

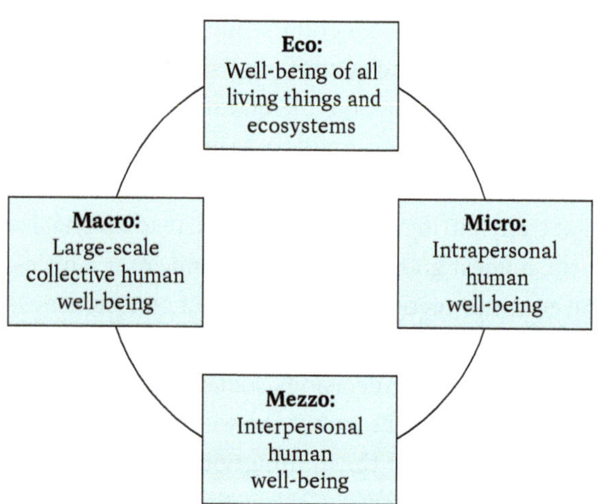

FIGURE 1.2 Micro, Mezzo, Macro, and Eco Levels of Practice

Inclusive Communities

You cannot co-create an inclusive community by excluding people, and you cannot have civilization without civility. When helping professionals engage in inclusive practice, we strive to welcome everyone, regardless of their diverse characteristics. However, an inclusive community survives and prospers when members agree on a code of civility that we label the "dialogue ground rules" in this text. These ground rules will be explored when we look at the use of dialogue in later chapters.

Responsibility

Inclusive practice emphasizes the responsibility every human being has for the well-being of themselves, other people, other living things, and the ecosystems that support all life. *Responsibility* is understood in this text as the "ability to respond." From that perspective, we are all responsible for the well-being of all living things and ecosystems. Each of us have the ability to respond and contribute something positive to the world, and each person's contribution is unique and equally valued.

Responsibility is good not only for our world, but also for ourselves. As we will discuss more in the text, when we human beings take actions for the highest good, there are ecobiopsychosocialspiritual benefits for us and for the world. Taking such positive actions can, for example, give us a sense of self-empowerment, purpose, and meaning in a world that may often seem frightening and senseless. Caring for other living things though such activities as community gardening, urban rewilding, stream cleanup, or promoting bicycle and pedestrian traffic can help promote our own physical and psychological well-being.

Faith-Based Activism

As Peter Frankopan (2023) observed, much of the history of the human species is about our struggle to survive in often hostile environments. Our ancestors have survived such dire challenges as ecosystems full of predators that want to eat us as well as ice ages, droughts, volcanic eruptions, hunting failures, crop failures, underpopulation, and overpopulation. See Science Resources Box 1.1 for a brief explanation of a human near-extinction event in the past.

SCIENCE RESOURCES 1.1 A prior near-extinction event associated with climate change

Although our current climate crisis is largely caused by human activity, our ancestors have apparently had to deal with "naturally" caused climate change throughout history.

During what is called the *Middle Pleistocene transition*, about 900,000 years ago, global climate went through a harsh cold climate event. Up until recently, scientists could not collect evidence from the scanty archeological records that have been found from that time. However, a new method of analyzing genetic codes called *FitCoal* (fast infinitesimal time coalescent process) was used by scientists to determine ancient human genetic changes.

Using this new method, scientists discovered that our ancestors went through a profound "population bottleneck" that began about 930,000 years ago and continued for about 120,000 years. There is evidence that global populations of our ancestors dropped to near-extinction levels at approximately 1,300 breeding individuals (Ashton & Stringer, 2023).

Today, of course, we face another hostile environment that this time is largely of our own making. We face the consequences of our technologies that continue to make such challenges as war, climate change, and pandemic ever more dangerous.

In our text, we will explore the limitations of despair and hope and propose the alternative of faith-based activism.

QUESTIONS FOR REFLECTION

1. In this chapter, we introduce key elements of inclusive practice. What are your emotional and cognitive responses to these ideas? You probably have preexisting feelings or beliefs about assessment and intervention. What are your responses to ecobiopsychosocialspiritual assessment, for example, or the four ways of knowing? How do you react to the idea of the paradigms of practice?
2. We also introduce the idea of the *radical middle*. How does this concept sit with you? Can you think of an example of a radical middle position that appeals to you, perhaps in such areas as politics, religion, or culture in which people currently often disagree?
3. The last section of this chapter is entitled "Other Principles in Inclusive Practice." Please pick one or two topics from this section and write a brief summary of what you think and how you feel about them.

REFERENCES

Ashton, N., & Stringer, C. (2023). Did our ancestors nearly die out? *Science, 381*(6661), 947–948. https://www.science.org/doi/10.1126/science.adj9484

Derezotes, D. S. (2023). *Inclusive social work: A new vision of community practice.* Cognella.

Derezotes, D. S. (2000). *Advanced Generalist Social Work Practice.* Sage Publications.

Duncan, B. L., Miller, S. D., Wampold, B. E., & Hubble, M. A. (2010). *The heart and soul of change: Delivering what works in therapy* (2nd ed.). American Psychological Association.

Frankopan, P. (2023). *The Earth transformed: An untold story.* Knopf.

Gaye, M. (1971). Mercy mercy me (the ecology) [Song]. On *What's going on.* Motown Records Corp.

McGilchrist, I. (2009). *The master and his emissary: The divided brain and the making of the Western world.* Yale University Press.

Mitchell, Joni. (1970). Big yellow taxi [Song]. On *Ladies of the canyon.* Reprise Records.

United Nations. (n.d.). *What is climate change?* Climate Action. https://www.un.org/en/climatechange/what-is-climate-change

Credits

IMG 2.1

Reconnecting With Our Inner Nature

Healing Body, Mind, Spirit

People will do anything, no matter how absurd, in order to avoid facing their own souls. One does not become enlightened by imagining figures of light, but by making the darkness conscious.

—CARL JUNG, *PSYCHOLOGY AND ALCHEMY* (1968) P. 99

Life's most persistent and urgent question is, What are you doing for others?

—DR. MARTIN LUTHER KING, JR. (SALTOS, 2015)

Today I started working on Chapter 2. It's a beautiful August morning, and during my bicycle commute down to Sugarhouse this morning, I suddenly smelled a familiar and sickening odor in the air and saw a young family in the driveway. It was time to take the children to school, but the lawn treatment truck had also arrived.

Trying to hold my breath as I pedaled past, I saw the father talking in the yard with the lawn treatment professional, who was spraying the grass as the children and family dog played nearby. The mother was opening the door to her van and calling to the children.

As I waved to them, I wondered whether they were aware that there are at least 40 common lawn and landscape chemicals that are potentially dangerous to people, pets, and the environment (including wildlife), with potentially serious short- and long-term effects (Beyond Pesticides, 2023). Perhaps they were not worried at least in part because the sides of the service truck bore a large slogan suggesting that the chemicals being sprayed on our soil and air and eventually reaching our waterways were completely safe and even healthy.

Perhaps this particular company, like growing numbers of other responsible firms, does not use dangerous chemicals. However, we know that more than 60,000 chemicals have been allowed on the market without any safety testing. The 2016 Frank R. Lautenberg Chemical Safety for the 21st Century Act does provide a means of testing and regulating chemicals, but experts estimate that our government scientists may need many decades or even centuries at current testing rates to examine the safety of these existing chemicals adequately. Unfortunately, political opposition to government regulation of our industries slows the process (Scialla, 2016; Environmental Protection Agency, 2023).

Regrettably, in perhaps most neighborhoods today, many of us contribute unnecessarily and without much awareness to the accumulation of greenhouse gases and other toxins in our environment. I watched the children as they boarded the big family pickup truck for their ride to the neighborhood school, which was only about three blocks away; their truck had an estimated gasoline mileage of 14 miles per gallon. Down the street, a neighbor was blowing the grass he just mowed off his sidewalk; his air blower and lawnmower had no pollution control devices on them. We know that gasoline-powered lawn and garden equipment currently emits about 26.7 million tons of pollutants each year in the United States (Banks & McConnell, 2011).

The Energy Policy Institute at the University of Chicago (EPIC) happened to release its annual Air Quality Life Index (AQLI) for 2023 this week, proclaiming that air pollution is now the biggest threat to global health. They estimate that more than 97% of the people on Earth today live in locations where air pollution exceeds recommended healthy levels, which shortens our average life span by an average of 2.2 years (EPIC, 2023). Also this week, several news networks covered concerns about the quality of the postepidemic indoor air our school children breathe in schools. We now know that poor-quality air exists in many of our nation's schools and that it has negative impacts on student learning and health (Ordway, 2023).

Most of the obstacles to local and global progress in developing and implementing climate solutions are inside of us. Our lack of consciousness about the nature of our world and ourselves contributes to our inability to work cooperatively with each other to face the challenges we live with today, including the climate crisis. How can I connect with nature if I cannot connect with my own human nature? Let's look further and explore why this loss of consciousness and connection has happened and what we helping professionals can do about it.

Developing a Climate Change in Human Consciousness

Two hundred years ago, naturalist, philosopher, and explorer Alexander von Humboldt (1769–1859) explained that "the most dangerous worldview is the worldview of those who have not viewed the world" and added that "there are three stages of scientific discovery: first, people deny it is true; then they deny it is important; finally, they credit the wrong person" (Goodreads, 2024). Humboldt's life and work inspired generations of naturalists, including early environmentalists such as George Marsh, Henry David Thoreau, and John Muir.

Humboldt's words are still relevant today. In this chapter, we explore the premise that we human beings have grown disconnected from our own bodies, minds, and spirits and that this disconnection is associated with our disconnection with the natural environment from which we came and where we live. We echo the words of Humboldt that many of us "have not viewed the world." We also see again that, regarding the scientific discovery of the climate crisis, many of us first "deny it is true; then they deny it is important" and will credit (i.e., listen to) only those who are in our own political group.

We have stated that human consciousness involves seeing the world the way it actually is, with clarity and acceptance. We know that more than 75% of the U.S. population now believe humans are causing climate change. However, there is still a large minority of the population (25%) who still may deny that climate change exists and that we humans are responsible. Part of what is going on is that the fossil fuel industry, political lobbyists, media moguls, and others have spread disinformation and misinformation to discourage the population to believe that the climate crisis is real; the five largest publicly owned gas and oil companies are estimated to spend about $200 million annually to delay or block effective climate solution policy (Maslin, 2019).

Table 2.1 describes five main reasons people continue to deny climate change, which were developed by professor of Earth science systems Mark Maslin (2019) at University College London. In column 2, the misinformation and disinformation used to deny climate change is briefly summarized, and in column 3, the scientific facts that dispute each of the five reasons are summarized. I put this information into a simplified table so that helping professionals can share it with clients, students, patients, and other community members as appropriate and use the table to help inform their practice.

TABLE 2.1 Five Main Reasons People Continue to Deny Climate Change

	False Argument From Climate Deniers	Science-Based Facts
1. Science denial	The science of climate crisis is not yet settled. Climate change is caused by nature, not humans.	For more than 30 years, the climate models that predict global temperature have shown consistently that there is a mostly human-caused climate crisis.
2. Economic denial	Climate change is too expensive to fix.	Current estimated cost of necessary climate solutions is only 1% of global gross domestic product (GDP). If we wait, costs will increase. U.S. subsidies for fossil fuel industry are about 6% of global GDP.
3. Humanitarian denial	Climate change is good for us, in part because warmer weather benefits farmers.	U.S. heat-related deaths are four times higher than cold-related deaths. Forty percent of the global human population live in tropics, which need no warming.
4. Political denial	Other countries are not taking action, so why should we?	Climate solutions are win-win solutions. Almost half of the total amount of carbon dioxide (CO_2) emissions comes from the United States and Europe.
5. Crisis denial	Climate change is not as bad as the scientists say. We can afford to wait.	A minority of humanity with the most privilege and power are the most immune from climate. They also tend to contribute more to harmful CO_2 emissions.

Source: Adapted from Maslin, 2019.

However, as most of us know, the facts themselves do not necessarily bring about a climate change in human consciousness. Several years ago, during the pandemic, a good friend of mine began texting me videos and articles that he had found that used the five arguments in Table 1.1. When the pandemic began to ease up and we were able to meet in person more safely, I asked him why he was texting me. He said he wanted me to have the "facts," and I responded that I could show him that those facts were, in fact, misinformation. But when he responded by saying, "And I can show you that they are true," I realized that it would not change anything for me to try to send him scientific reviews because he was not open to changing his mind. It was best to continue to care about my friend and wish him well, since loving relationships *can* bring about culture change. Probably, many of us are like my friend; up to one-quarter of our fellow citizens, or about 85 million people, may be unwilling to change their minds because of scientific research alone.

Why do we humans so often become closed off to seeing things different? Professor Florian Zimmermann, an economist at the University of Bonn, has found that the denial of human-made global warming may often be part of "the political identity of certain groups of people." Because they identify with their political beliefs, they may be unlikely to be influenced by the results of scientific research (University of Bonn, 2024).

A climate change in human consciousness is unlikely to be brought about merely by the dissemination of scientific evidence. How can professional helpers assist their community members to make the climate changes necessary for us to cooperate together in designing and implementing sound climate solutions? Professor Zimmerman's finding that identification with political party may be a primary reason for resistance to scientific evidence suggests where we may shift our focus.

In our introduction, we defined *ego* as that part of our mind that identifies with form, especially mental forms or beliefs. As Eckhart Tolle (2005) suggests, consciousness is the tool we have to transform identification. And, according to Tolle, when I am aware of my ego and my identifications, they no longer have power over me. In our culture, most of us identify with our bodies, for example, thinking that our bodies are us and that we are our bodies. However, such identification is unfulfilling because all bodies are constantly changing and aging and eventually die. We also identify with our minds, thinking that they are us and that we are them. This identification is also unfulfilling, because the mind is constantly chattering with the collection of often unhelpful concerns that we call anxiety.

And why do we humans so often seek to identify with political parties or with other group's belief system? Remember that identification goes beyond mere membership in a group; when I identify, I see my beliefs or possession as actually being *me*, or at least a part of me. Renowned spiritual teacher J. Krishnamurti (1996) taught that we seek attachment through identification because we fear the lonely separateness and emptiness of our current existence. However, he adds, such identification ultimately lead to additional suffering because our loss of connection with the world and ourselves actually further worsens when we try to substitute identification with connection.

When we identify with our beliefs, we do not want to examine whether they are true. When we are confronted with the painful truth of the climate crisis, we tend to react psychologically in one (or both) of two ways that at first seem to represent opposite extremes.

The first reaction is associated with despair, helplessness, and retreat. Now that a majority (more than 60%) of people in the United States see global climate change as a "major threat," we may hear that people are suffering from such syndromes as "climate anxiety," "climate dread," and "climate fatigue" (Ruiz-Grossman,

2021). For example, when I work with colleagues who teach environmental studies classes, a number of them are often concerned that the curriculum was causing students harm by apparently increasing their depression and anxiety and despair.

The second and opposing reaction is what Mishra (2017) calls *resentiment*, which he saw as a chronic state of anger and hostility toward other populations who are typically perceived as being more privileged than us and as being oppressive toward us. There is evidence that many people are irritated by environmentalists and other activists because they seem aggressive, militant, and unconventional (Shire, 2015). Additionally, about 13% of Americans polled in a survey of 23 nations conducted by the YouGov-Cambridge Globalism project believe that our global climate is changing, "but human activity is not responsible at all." And another 5% said our climate is not changing at all. These statistics show that the United States has the highest percentage of people who identify as climate deniers in the wealthy world (Milman & Harvey, 2019).

These two types of responses to the climate crisis can be positioned at opposite ends of a continuum, as illustrated in Figure 2.1. As you can see, the "passive" response is associated in the figure with helplessness, self-blame, and general withdrawal from the community. On the opposite end of the continuum is what we call the "aggressive" response, which is associated with an attitude of hostility and blaming and attacking others (through words and/or deeds).

In the "radical middle" between these two extreme reactions is what we call "assertive," which is associated with such traits as consciousness, faith-based activism, and a sense of responsibility and service to community. Assertive responses are based on seeing things the way they actually are (inside myself and in the world) as well as with a sense of responsibility for the welfare of other people, other living things, and the ecosystems that keep us all alive. We will also look more in depth at the concept of faith-based activism in this chapter as well as later in our text.

We can now further explore the psychology of these three positions and then describe some approaches to healing and transformation of consciousness. The passive and aggressive opposites can actually be understood as representing two sides of the same coin. This is because these two extremes share some common underlying psychological traits. Both typically use the defenses of minimization, denial, and avoidance of the hard reality of the climate crisis. Both passive and aggressive responses are also associated with what McGilchrist (2009) called the current tendency in Western culture for people to feel both "impotence and moral superiority." *Impotence* here refers to a belief that I am unable to do much of anything to meaningfully change my world for the better. *Moral superiority* here means that regardless of whether or not I feel empowered enough to try to change the world, I nevertheless think that my beliefs and actions are based on values superior to those held by those who disagree with me, and therefore, I think I am justified in any thoughts and behaviors I might direct toward them.

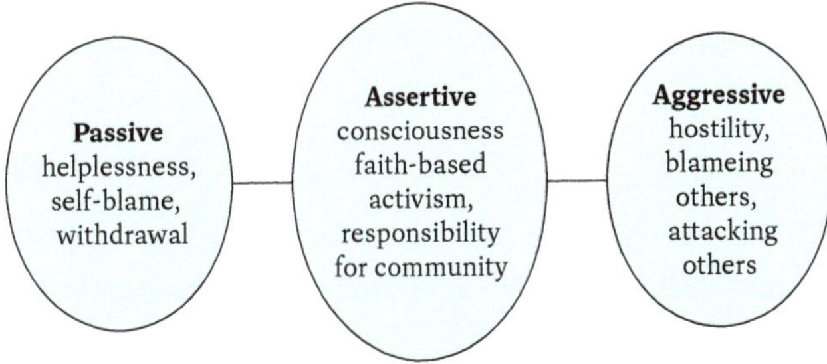

FIGURE 2.1 A Continuum of Our Responses to the Climate Crisis

Passive and aggressive responses are also both associated with the symptoms of anxiety and depression that are increasingly seen in our young people. As we have seen, in the U.S. Surgeon General's Advisory (Office of the Surgeon General, 2021), climate change is described one of the macro-level factors that can affect the mental health of young people, including anxiety and depression. Similar associations between macro-level factors and individual well-being have been found with adolescents living in many other countries across the world (Currie & Morgan, 2020).

Healing and Transformational Work

What are some practical ways that helping professionals can engage, assess, and intervene with our clients during the climate crisis to assist them in developing consciousness and connection? We can explore this question by considering all the ecobiopsychosocialspiritual dimensions of human development and see what emotional, physical, cognitive, social, and spiritual healing and transformation might look like today. We will start by discussing emotional healing in depth and then building on that discussion with the following four dimensions.

Emotional Healing and Transformation

What does it mean then to heal our emotional responses to the climate crisis? In order to assist them with their emotional health, helping professionals essentially need to help our clients become conscious of their emotions, which means becoming aware and accepting of them. As people grow more conscious of their emotions, we can connect with our world emotionally.

We want to help our clients develop their emotional intelligence, which includes developing the critical feeling necessary to determine what are emotional responses are. We encourage our clients to see that our emotions *and* feelings are always subjective and therefore must be examined carefully, using all the ways of knowing. Critical feeling does not replace the critical thinking currently emphasized in higher education but adds to critical thinking. We know that as our clients develop emotional consciousness, they can also learn how to avoid unnecessary and unhelpful reactivity to the climate crisis and also how to better judge whether what they read or hear is misinformation or disinformation (which is intentionally shared misinformation).

Emotional and transformation involves becoming conscious of both our core *emotional* responses as well as of the *feelings* we often create over time by thinking about those emotions. What is the relationship between emotions and feelings? In our text, we will define emotions and feelings as interrelated but different subjective human experiences. *Emotions* are short-term responses to events and situations that can influence our thoughts and behaviors. We can teach that there are maybe six primary emotions—sad, mad, glad, scared, excited, and disgusted—just like there are three primary colors—red, blue, and yellow. In contrast, *feelings* can be described as neocortical-brain reactions to our emotions, in which we give meaning to our emotions. A simple way to summarize all of this is to say that feelings are the result of the interpretations we give our emotions, as summarized by the following equation:

Emotions + thoughts = feelings

For example, if a dog chases me on my bike and tries to bite me, my first and primary emotion might be fear during the brief 30 seconds that I am attacked. Then, later, as I pedal away, I might start to think about what happened and say to myself, "I am outraged! What a stupid dog ... and where was the owner when this happened?"

Indeed, I might start to feel increasing outrage all day toward the dog and its owner, and when I see them on my ride later that evening, I might want to stop and scream, "You are an irresponsible dog and dog owner!"

I have added my interpretation to my initial emotion of fear and created outrage, which is a self-righteous and resentful anger.

As Krishnamurti (1996) reminds us, all thoughts arise from past conditioning and experiences, and knowledge derived from past conditioning and experiences is by its nature always limited. However, when we are conscious of the limitations of our thinking, it becomes possible to begin to see the world without our thinking getting in the way and thus observe ourselves and our world the way it actually is. This consciousness could be said to be the goal of what we now call *meditation* and *mindfulness*. This goal does not mean that the mind is viewed as useless, of course, but rather that it has two equally important functions: to think and to be quiet. A quiet mind allows me to see myself and the world without being necessarily bound by past conditioning; I can be thus be conscious by using all the ways of knowing available to humans.

What emotional responses do we typically have today in our reaction to our accelerating climate crisis? To answer this question, first, ask yourself what kinds of emotions you feel. I have personally observed a wide range of emotions in myself but found that the most common core emotions seem to be sadness, fear, and anger.

As professional helpers, we want to help our clients develop and skillfully use their emotional intelligence, which is their ability to notice their emotions and feelings, and then communicate and act on them effectively. In Figure 1.1, the assertive, "radical middle" position represents this ability. Rather than suppressing my emotions and feelings and attacking myself (the passive position) or striving to suppress or attack the emotions and feelings of other people (the aggressive position), the assertive person is able to express their emotions and feelings directly without attacking anyone.

For example, returning to the dog attack story we looked at earlier, a passive reaction to the dog owner might be to say nothing at all and instead to beat myself up for being bitten and being upset about it. An aggressive response might be to scream curse words at the owner as I try to kick their dog. An assertive response would be to stop and say hello to the owner and then explain directly but calmly what happened when their dog attacked me earlier that morning. Professional helpers can offer such examples to their clients when teaching them how to develop their emotional intelligence.

We help our clients see and accept their passive and/or aggressive patterns. We acknowledge with them that it is not pleasant to have to face such difficult emotional responses as fear, sadness, anger and helplessness as we increasingly hear more bad news about the climate crisis. We may also point out to them something most people already know—that when we minimize or deny emotions, they do not go away. When we human beings deny our emotions, we do not ultimately avoid pain but simply substitute such symptoms as anxiety and depression, powerlessness and moral superiority, passivity, or aggression for the core emotions of fear, sadness, and anger we may be having.

When we are working with people who are capable of seeing and willing to see their own psychological dynamics, we can also explore with them the role left-brain thinking plays in turning our emotional responses into unhelpful and harmful feelings. Our increasing left-hemispheric-brain dominance and related egoic dominance support such feelings as impotence and moral superiority, passivity and aggression. As we have discussed, McGilchrist (2009) has shown evidence that left-brain functioning has progressively dominated Western culture over many centuries, resulting such additional egoic symptoms as identification with personal beliefs, a sense of personal entitlement and specialness, a focus on short-term personal gain and security, and lack of compassion for others who seem different.

We also want to help our clients appreciate that emotions and feelings are fascinating and complex processes that we can continue to work on understanding and expressing, across our lifespans. For example, a complex interaction between the emotions and feelings we may have about climate change today may well support the experience of powerlessness that we see so often today. TenHouten (2016) linked powerlessness with the four "primary emotions" of "acceptance-acquiescence," "anticipation-expectation," sadness, and fear as well as with the six "secondary-level" emotions of fatalism, pessimism, resignation, anxiety, submissiveness, and shame. These are all emotions that we may hear people talk about and express today in relation to the climate crisis.

We helping professionals also remind our clients (and ourselves!) that the goal of emotional healing and transformation is not to escape from emotional pain or inflict pain on self or others, but to understand and accept our emotions so that we can express them effectively for the highest good. This insight is related to the principles about fixing and healing that we introduced in Chapter 1. Emotional pain can be healed (made whole) but not fixed (eliminated). We also can suggest that emotional pain can get our attention and lead to increased consciousness.

We also help our clients understand that emotions can provide us with valuable data about what is going on with world and ourselves. We also emphasize that our feelings (interpreted emotions) can also give us valuable insight, particularly into how our own experiences and conditioning influence our interpretation of our emotions. For example, as we said earlier, as a person becomes more conscious of emotions and feelings, they can become better able to judge whether what they read or hear is misinformation or disinformation. As we become less susceptible to having our feelings become triggered and overwhelming, we can analyze new ideas and information more objectively.

Our consciousness of our emotions and feelings allows us to actually grieve the losses and anticipatory losses we are experiencing during the climate crisis. The work of loss is grief, which means that I let myself feel the various emotions and feelings I may have as I become ready to do so. We know that the grieving process varies in different people and that there not as many *stages* of grief that we all undergo in a tidy order as there are *fluctuations*, which may be experiences in waves of various intensity (Gard, 2020).

Helping professionals can normalize the complicated grief that many of our clients experience every day related to the climate crisis. These complications are the complex mix of emotions and feelings humans naturally have when we experience such losses as the passing away of clean water, earth, and air; living species; relatively undeveloped green spaces; of predictable climate; steady ocean levels; and a sense of security and prosperity. Depending on such factors as each person's socioeconomic status, political power, and cultural-racial privileges, these losses can vary in intensity and scope.

Helping professionals need to do our own work in seeing how we grieve our own personal losses and to understand generally how and why other people trigger our own emotional reactions. Whether we personally trend more toward the position of passivity or aggression, our own patterns will influence the way we view and react to the processes of our clients. More specifically, our own emotions and feelings about the climate crisis can also influence how we react to the ways in which our clients react to the crisis. We may have what we could call *climate countertransference*. For example, if I am a counselor who is also a strong climate change denier and have a client who is an enthusiastic climate change activist, I may become intensely judgmental toward the client.

Although we helping professionals have countertransference reactions to every person with whom we work, sometimes those reactions are particularly intense and uncomfortable. When such intensity appears, it often mean that I have some work to do in exploring why I was so strongly triggered. In general, the questions I teach my students to consider regarding countertransference reactions they may have include the following:

1. How possible is it that the way I feel toward and think about the client/student/patient is the way most other people in the client's life also feel and think? And if so, what does that tell me about the client?
2. How possible is it that the way I feel toward and think about the client/student/patient is the way the client feels and thinks about themself? And if so, what does that tell me about the client?

For example, if I am usually annoyed with the way the client is rigid and self-righteous in their thinking about climate, then maybe most people also feel that way. This would probably result in the client feeling rejected and lonely since they are probably often rejected and disliked in their everyday life. Similarly, the client also probably is likely to not have much positive self-esteem, which would likely further increase their suffering. Therefore, I can use my countertransference reactions to help me better understand and even have more compassion for the people with whom I work.

Physical Healing and Transformation

In Chip Ward's essay "Landing Under a Sleeping Rainbow," he writes about how he and his family became more aware of how they had learned to disconnect their own bodies from nature:

> During our years in the redrock wildlands, we learned compelling and fundamental lessons that we had missed during our modern education while immersed in the American way. For the first time in our lives, the connections between our bodies and the water and soil that nourished us appeared short and simple. ... Our bodies are a community of fluids. While living on the ranch, the ways water reached our habitat from the sky and how we incorporated it became obvious. (Ward, 2019, pp. 183–184)

Although we all have bodies, of course, most of us are not aware of where and how our food and water is produced, nor are we typically cognizant of what happens to our waste after we flush the toilet. As Ward suggests, we tend to simply pay our grocery, water, and sewer bills and perhaps grumble about how much they cost.

In addition, we are often ourselves disconnected from our own bodies. British psychoanalyst Susan Orbach's (2019) book *Bodies* describes the unhappiness and even "body hatred" perhaps most people in Western countries as well as many other locations on the planet have with our own physiques. She notes how often we see advertisements that try to make us think that we need help, for example, reducing our weight, reversing the aging process, sustaining youthful-looking skin, or increasing our athletic or sexual performance.

This current attitude toward our bodies is certainly not a result of conscious, reverent awareness, but actually much more the opposite. We do not tend to see our bodies accurately nor do we accept them for what they are. Instead, as Orbach suggests, we tend to have unrealistic expectations for our body appearance and body functioning and therefore tend to see our bodies from a perspective of deficit, guilt, and shame.

Although we human beings are animals, we often speak in everyday language as if humans are not part of the animal kingdom; for example, we might say, "I think human beings are more important than animals" as if we were not also animals. The approximately 8.7 million species living on Earth today are divided by scientists into five kingdoms: animalia, plantae, fungi, protista, and monera. We humans are classified as belonging to the animalia kingdom. Scientific facts can help introduce to our clients, patients, and students to the deep connection we have with other life forms. For example, a group of scientists from Israel and Canada estimated that we humans have approximately an equal number of bacterial and human cells in our bodies (Davies, 2018).

Physical healing and transformation during the climate crisis thus requires a consciousness about how we feel, think about, and interact with our bodies. When I understand the extent to which I do not see my true nature and how my conditioning and experiences have resulted in my perceptions about my body, then I can begin to see what the true nature of my body really is and how my body is intimately connected with Earth and other living things. We can suggest to our clients that they explore these issues for themselves. I find that yoga work, for example, can help me become more conscious of my body, and many of my students and clients agree.

One of my favorite ways to approach this work is to suggest that my clients and students "think like a plant does" or "think like another animal does" about their bodies. For example, I might ask them if they think that a pine tree hates the way they look. Or, I might ask if they think a butterfly worries about their attractiveness. We then can start to talk about why we human animals have learned to disconnect from and even hate our bodies. Helping professionals realize, of course, that body image will vary with such factors as gender identification, race, culture, and socioeconomic level. I also like to give people who are out of touch with their bodies and seldom visit more natural environments, an assignment to go out in nature—by themselves, if it can be done safely—and then report later about what they experienced and learned.

Cognitive Healing and Transformation

When I was a child growing up in Chicago, we had three news networks that we could listen to. Most people heard the same information about the world, so our conversations were mostly based on a common shared experience, even if we did not always agree on our interpretation of the facts. Today, we obtain our news through multiple sources, including television, computers, cell phones, radio, printed magazines and newspapers, tablets, and various electronics that link Internet news with our television. In 2014, Americans typically used four different devices or technologies each week; about 87% of us use television, 69% computers, 65% radio, and 61% magazines and newspapers (American Press Institute, 2014). Less than a decade later, about 86% were getting news from a variety of digital devices, including smartphones, tablets, and computers (Shearer, 2021). In addition, there are now many news sources that obviously cater to a particular political viewpoint to an extent that we did not see decades ago.

What are cognitive healing and transformation in relation to the climate crisis? In an era of multiple and conflicting news sources, a significant part of cognitive healing and transformation involves the use of critical thinking to understand the extent to what we hear and currently believe is actually misinformation or disinformation. The Science Resources 2.1 box has three online sources that provide useful and practical advice on how to detect misinformation and disinformation from which clients may benefit.

SCIENCE RESOURCES 2.1 Detecting misinformation or disinformation

Following are three online sources for advice on how to detect misinformation and disinformation that my students have found useful:

1. Ancis, J. R. (2021). Disinformation techniques: How to spot them. *Psychology Today*, July 15, 2021. https://www.psychologytoday.com/us/blog/the-cyberpsychology-page/202107/disinformation-techniques-how-spot-them
2. Baig, E. C. (2022). *Ten ways to spot fake videos and falsehoods on the Internet*. AARP, August 19, 2021. https://www.aarp.org/home-family/personal-technology/info-2022/identifying-disinformation.html
3. Union of Concerned Scientists. (2022). *How to spot disinformation*. April 1, 2022.https://www.ucsusa.org/resources/how-spot-disinformation

Questions that these sites suggest asking include the following:

- Who are the authors?
- What evidence is presented? What is the original source?
- Do other sources agree?
- Is it hard to separate facts from opinions?
- Are experts from reputable organizations cited?
- Does it confirm my beliefs or appeal to my emotions?
- Does it offer any nuance, or is everything presented in black and white?

Social Healing and Transformation

I have frequently seen that a climate change in consciousness can happen through relationship. Each of us has at least some influence over the people with whom we interact in our lives. Relational influence may be stronger than ever in this time of aloneness. We could say that the loneliness and isolation in our communities is one of our biggest social challenges today (U.S. Department of Health and Human Services, 2023). The way I like to describe this issue is that, if I had a magic wand today, the first thing I would wish for all the people in our country is that we all belong to an *inclusive community.* We could start to define such a community as having the following characteristics:

1. Everyone is welcome.
2. We are all responsible for the community's well-being.
3. All voices are heard and contributions seen as we seek consensus in decision-making.
4. There are agreed-upon rules of civil engagement, dialogue, and conflict resolution.
5. We value our common humanness as well as our individual differences.
6. We value the common good but can grant amnesty when people make mistakes.

When we look for the etiologies of this epidemic, we can reflect on one of the foundational ideas of our nation. I believe that the phrase *pursuit of happiness* has been misunderstood. As Emory University (2018) points out, according to the Declaration of Independence, "the extended quality of happiness—what we might call the good or flourishing life—is or should be a primary concern of government. That means it isn't just about *my* happiness, especially idiosyncratically defined, but about *all citizens'* happiness" (p. 1). Perhaps, indeed, we have overemphasized our individual independence and importance and forgotten that true happiness is about love for others, including other people, other living things, and our ecosystems.

What, then, is social healing and transformation? Perhaps it involves, at least in large part, the transformation of our families, institutions, villages, states, and nations into inclusive communities. In order to co-create such inclusive communities, we helping professionals want to help our clients become conscious of what kinds of social connections they need and have.

An examination by Mike Glauser (2022) of individual happiness from religious wisdom traditions, philosophers, and modern science generated six principles of happiness. These include giving up ego, refraining from judging, doing good deeds daily, forgiving others, sharing one's good fortune, and caring for our needy. All these principles share one common value—the idea that human happiness comes more from a focus on the well-being of communities than on our individual well-being.

Spiritual Healing and Transformation

Finally, what is spiritual healing and transformation, in a time of climate crisis? Whereas *religiosity* can be conceptualized as a *social* dimension in which we humans share rituals doctrines and beliefs within a community, *spirituality* can be defined as an *individual* dimension of human development that has to do with developing consciousness of our connection with everything in the universe. Thus, a person can consider themselves to be either religious, spiritual, both, or neither.

Perhaps a significant part of the epidemic of loneliness and isolation that we discussed earlier is related to our attempts to fill spiritual need with our social relationships with other people. Although there is little doubt that most humans have a need for inclusive communities in our lives, we could also say that no human relationship can alone fill the need for spiritual connection that most humans also have. Perhaps we tend to believe today that if we find the right partner, our romantic love might fill that spiritual need, or perhaps we think the same things about creating the perfect children who will fulfill our life needs, or perhaps the perfect job that will make everything all right. But as Glauser (2022) points out, our wisdom traditions deny that human relationships, wealth, power, or social status will by themselves make us happy.

One spiritual practice that my students and clients have found useful in dealing with the climate crisis and other micro-, mezzo-, micro-, and eco-level issues is dying every day. Although this title at first may sound depressing and even morbid, the practice can be quite healing and uplifting. The idea is based on the insight that death can be viewed as the process of letting go of what is not me. What does that insight actually mean? Each person may imagine the answer very differently, but most people can make a list of things they identify with that are not them (see Table 2.2).

TABLE 2.2 Possible List of "Things That I Identify With That Are Actually Not Me"

Things I Identify With	Specific Examples
My possessions	House, car, clothing, books, works of art, money, gold coins
My social roles	Grandparent, parent, child, elder, personal fame, power over others
My titles	Boss, director, promotion, academic degree, profession, job
My beliefs	Religion, nationality, political party, personal philosophy
My body	Youthful appearance, athletic ability, gender, sexual activity

People can be asked to make such a list and reflect on the ones with which they most strongly identify. Remember that we said that *identification* is the basic function of the ego and involves taking the attitude that "that is me and I am that." When we can take on the challenge of disidentification, I become conscious of the fact that I am "more" than any of those things with which I have previously identified. I do not actually have to give up my house or vocation, for example, to disidentify with those parts of me, but I can develop a different, more detached relationship with them. My consciousness of my identification thus helps me disidentify from those things to which I am most attached.

What does disidentification have to do with the climate crisis? We could say that disidentification work helps us free ourselves from the influence of our ego. As we become less identified with the things we personally "own," such as material goods, roles, titles, beliefs, and physical attributes, we may become more able to value those things we do not own but can value and enjoy, such as other people, other living things, and the ecosystems that support all life. We may become, in fact, aware of our connection with everything.

What, then, does it mean to be conscious of my connection with everything? Perhaps this is a question that cannot answered in any textbook. As Krishnamurti (1996) has suggested, the one who says they understand such things probably does not. Yet, most people I know have reported that they have been able to actually feel that connection in their hearts, at least from time to time.

QUESTIONS FOR REFLECTION

1. For most helping professionals, our most challenging clients are often the ones with whom we have a difficult countertransference. What types of clients are likely to be the most difficult for you? For example, how do you expect that you would react to clients who are climate change believers or deniers? Please explain.
2. When we assess our clients, we need to be able to assess the extent to which they may need more healing. Which of the five dimensions (emotional, physical, cognitive, social, spiritual) do you personally think are the healthiest within yourself? Which require the most healing and transformation?
3. If you were your own therapist or doctor, what methods would you advise yourself to take, to start the healing work you most need today, in this time of climate crisis?
4. Have you ever felt a connection with everything in the universe? Please explain your answer.
5. What items in Table 2.2 are you most attached to today? What might it take to let go of these things?

REFERENCES

American Press Institute. (2014). *How Americans get their news.* https://www.americanpressinstitute.org/publications/reports/survey-research/how-americans-get-news/

Ancis, J. R. (2021). Disinformation techniques: How to spot them. *Psychology Today,* July 15, 2021. https://www.psychologytoday.com/us/blog/the-cyberpsychology-page/202107/disinformation-techniques-how-spot-them

Banks, J. L., & McConnell. (2011). *National emissions from lawn and garden equipment.* Environmental Protection Agency. https://www.epa.gov/sites/default/files/2015-09/documents/banks.pdf

Beyond Pesticides. (2023). *40 common lawn and landscape chemicals.* https://www.beyondpesticides.org/resources/garden-pesticides

Currie, C., & Morgan, A. (2020). A bio-ecological framing of evidence on the determinants of adolescent mental health—A scoping review of the international Health Behaviour in School-Aged Children (HBSC) study 1983–2020. *SSM—Population Health, 21*(12), 100697. https://pubmed.ncbi.nlm.nih.gov/33335971/

Davies, E. (2018). *What proportion of the human body is bacteria (and how do we measure it)?* BBC Science Focus, July 7, 2018. https://www.sciencefocus.com/the-human-body/what-proportion-of-the-human-body-is-bacteria-and-how-do-we-measure-it

Emory University. (2018). *What the Declaration of Independence really means by "pursuit of happiness."* Emory News Center, July 3, 2018. https://news.emory.edu/stories/2014/06/er_pursuit_of_happiness/campus.html

Energy Police Institute at the University of Chicago. (2023). *Air pollution cuts life expectancy by more than two years, study says.* https://epic.uchicago.edu/news/air-pollution-cuts-life-expectancy-by-more-than-two-years-study-says/

Environmental Protection Agency. (2023). *Highlights of key provisions in the Frank R. Lautenberg Chemical Safety for the 21st Century Act.* https://www.epa.gov/assessing-and-managing-chemicals-under-tsca/highlights-key-provisions-frank-r-lautenberg-chemical

Gard, B. (2020). *Grief and loss: Understanding the process.* Psychology Today, December 21, 2020 https://www.psychologytoday.com/us/blog/psychological-trauma-coping-and-resilience/202012/grief-and-loss-understanding-the-process

Glauser, M. (2022). *One people one planet: Six universal truths for being happy together.* Lioncrest Publishing.

Goodreads. (2024). *Alexander von Humboldt quotes.* https://www.goodreads.com/author/quotes/303739.Alexander_von_Humboldt

Jung, C. (1968). *Psychology and Alchemy.* Princeton, N.J.: Princeton University Press.

Krishnamurti, J. (1996). *Total freedom: The essential Krishnamurti.* HarperOne.

Maslin, M. (2019). *Here are five of the main reasons people continue to deny climate change.* ScienceAlert, November 28, 2019. https://www.sciencealert.com/the-five-corrupt-pillars-of-climate-change-denial

McGilchrist, I. (2009). *The master and his emissary: The divided brain and the making of the Western world.* Yale University Press.

Milman, O., & Harvey, F. (2019). U.S. is hotbed of climate change denial, major global survey finds. *The Guardian*, May 8, 2019. https://www.theguardian.com/environment/2019/may/07/us-hotbed-climate-change-denial-international-poll

Mishra, P. (2017). *Age of anger: A history of the present.* Farrar, Straus, and Giroux.

Office of the Surgeon General. (2021). *Protecting youth mental health: The U.S. surgeon general's advisory.* https://www.hhs.gov/sites/default/files/surgeon-general-youth-mental-health-advisory.pdf

Orbach, S. (2019). *Bodies.* Profile Books.

Ordway, D.-M. (2023). How indoor air quality in schools affects student learning and health. *The Journalist's Resource*, April 12, 2023. Harvard Kennedy School's Shorenstein Center. https://journalistsresource.org/education/indoor-air-quality-schools-student-learning-health/

Ruiz-Grossman, S. (2021). *The antidote to climate dread.* HuffPost, August 25, 2021. https://www.huffpost.com/entry/climate-change-crisis-dread-fatigue-action_n_61268ecde4b0f562f3d9f07b

Saltos, G. L. (2015). *MLK, Jr., asked us, "What are you doing for others?" Here's how we answered.* HuffPost, January 19, 2015. https://www.huffpost.com/entry/mlk-day-serving-others_n_6489236

Scialla, M. (2016). *It could take centuries for EPA to test all the unregulated chemicals under a new landmark bill.* PBS News, June 22, 2016. https://www.pbs.org/newshour/science/it-could-take-centuries-for-epa-to-test-all-the-unregulated-chemicals-under-a-new-landmark-bill

Shearer, E. (2021). *More than eight in 10 Americans get news from digital devices.* Pew Research Center, January 12, 2021. https://www.pewresearch.org/short-reads/2021/01/12/more-than-eight-in-ten-americans-get-news-from-digital-devices/

Shire, E. (2015). *Why people hate feminists, environmentalists, and activists in general.* The Week, January 10, 2015. https://theweek.com/articles/459460/why-people-hate-feminists-environmentalists-activists-general

TenHouten, W. D. (2016). The emotions of powerlessness. *Journal of Political Power, 9*(1), 83–121.

Tolle, E. (2005). *A new earth: Awakening to your life's purpose.* Plume.

Union of Concerned Scientists. (2022). *How to spot disinformation.* April 1, 2022. https://www.ucsusa.org/resources/how-spot-disinformation

University of Bonn. (2024). *Why are people climate change deniers? Study reveals unexpected results.* February 2, 2024. https://phys.org/news/2024-02-people-climate-deniers-reveals-unexpected.html

U.S. Department of Health and Human Services. (2023). *New surgeon general advisory raises alarm about the devastating impact of the epidemic of loneliness and isolation in the United States.* https://www.hhs.gov/about/news/2023/05/03/new-surgeon-general-advisory-raises-alarm-about-devastating-impact-epidemic-loneliness-isolation-united-states.html

Ward, C. (2019). Chapter 26: Landing under a sleeping rainbow. In S. Trimble (Ed.), *The capital reef reader* (pp. 178–185). University of Utah Press.

Credits

IMG 3.1

Healing Our Ego-Hemispheric Imbalances

The attempt to escape from pain, is what creates more pain.

—GABOR MATÉ (MATÉ, N.D.)

The moment you become aware of the ego in you, it is strictly speaking no longer the ego, but just an old, conditioned mind-pattern. Ego implies unawareness. Awareness and ego cannot coexist.

—ECKHART TOLLE (GOODREADS, N.D.)

I remember when I was a young man sitting on a commuter train and observing everyone in the car on their way to work. As I recall, I suddenly began to realize, for some reason, how everyone on the train seemed captured by a sense of their own self-importance and therefore subjugated to their ego. Each person seemed to be carrying around a feeling of grandiosity through which they felt in many ways superior to the other people in the train. And if they also had feelings of inferiority, I realized that they probably imagined in their grandiosity that their suffering was also greater than anyone else's suffering.

Although I was correct in the assessment that most of us are dominated by our ego and perhaps have little if any awareness of our own interrelated grandiosity and inferiority, I was not yet aware of another even more important fact: that I am just like them! As a friend of mine shared with me after returning from a workshop in California, he had learned that when it comes to observing traits in other people, "if you spot it, you've probably got it."

And perhaps that is how most of us may begin to become aware of the ego—by first seeing it in others. Only later may we understand that ego seems to be a characteristic of human consciousness. Or, we could say, "If you have a brain, you have an ego."

Hopefully, I can now write the rest of this chapter about ego without letting my ego take too firm a hold on my own thinking. The ego can be clever, and I find that it can sneak into my mind at any time. However, fortunately, as we will see, we have an important tool we can use to temper the ego, which is our own consciousness, or reverent awareness.

Introducing the Ego

Consider whether it is true that you, like me, have a tendency to think about how your world "should be" and how you should be treated in that world, and therefore you experience endless conflict with how things actually are. This is a common symptom of that fundamental part of our mind we call the *ego*, along with such related symptoms as self-interest, grandiosity, and inferiority.

The Latin word for "I" is *ego*. We can think of the ego as identification with a structure or form, especially with our mental structures or beliefs (Tolle, 2006). What is an *identification*? When I see something as being the same as myself. For example, if I identify with a particular religious belief, group of people, nationality, or political affiliation, then I view the beliefs of that group or affiliation as an essential aspect of myself. Unfortunately, when I identify with anything, I may also feel justified in defending any perceived attack on them because such an attack feels like an attack on myself. Anger, resentment, and violence all become more likely.

In this chapter, we will examine how ego is especially related to our left-hemispheric brain functioning and how the apparently still-increasing dominance (McGilchrist, 2009) of that functioning over the right hemisphere is associated with our climate crisis. We will also review how our awareness of our ego and this overall left-hemispheric dominance can help us create a favorable climate change, both in our own consciousness and, ultimately, in our humansystems and ecosystems as well.

We use the word *imbalances* in the title of Chapter 3 to emphasize that the healing work is to *balance* different parts of our mind rather than to futilely attempt to discard our unwanted parts. We cannot destroy the ego, and the human ego and other associated left-hemispheric characteristics are potentially useful elements of human consciousness. We can, however, become aware of the ego and thus heal it.

Like any other part of our minds, the ego needs to be *tempered*, which means to find the "right mix at the right time." We will examine how as we become aware of our ego-hemispheric dominance, we can temper it and allow for other ways of thinking to also be felt and expressed, thus healing our consciousness and making it whole. Much like the story in the preface about the two wolves that must both be fed, we want to feed both hemispheres of our brain. We need them both.

Ego and Left-Hemispheric Dominance

The human ego, which can be expressed at both the individual and collective levels, probably evolved at times in human history when increased individuality and self-interest helped promote the survival and well-being of our ancestor groups. Ego appears to be part of the human mind. We are born with it, and it is tempered only when we become conscious of it, with reverent awareness.

For example, people with a strong self-interest might have had a survival advantage in a harsh environment such as Northern Eurasia during the last Ice Age, when people may have lived together in small family groups. The drive to look after only oneself and their immediate family might have helped them survive when few resources were available.

And small groups of pastoralists may have also benefited from enhanced self-interest because they also needed to work alone or in small groups while tending their sheep or cattle. In contrast, groups of farmers or fishermen may have benefited more from an enhanced collective interest in living in cultures in which people cooperate to do the work necessary to survive.

In our past, individual humans with the strongest ego may also have had an advantage in competition with other animals and perhaps especially with other hominins, including other members of our own species, extinct human species, and our immediate ancestors. In our own competitive modern world, contemporary humans with the most dominant ego may still compete more successfully for power, wealth, and fame because they are willing to do anything to succeed within the businesses, corporations, or institutions in which they work.

Indeed, I often wonder whether concentrated wealth and power may be two of the most important factors associated with the evolution of ego as well as with the worsening climate crisis this text addresses. Here, *concentrated wealth* refers to a chronic and unequal distribution of power and wealth between human groups. For example, when one culture starts to create surplus wealth, such as extra grains and dried fish that cannot all be eaten immediately, it needs to be stored, and the owners might also feel the need to protect it from other cultures with fewer resources. It is easy for those with greater resources and power to identify with what they have and to thus see themselves as wealthy, powerful, and superior people.

The human ego also may serve to justify our misbehavior toward other living things and the ecosystems that support all life. If I feel superior to other beings, I can also feel justified to do take whatever I want from nature and destroy whatever is left. The pursuit of wealth and power requires intense extraction of natural resources from our ecosystems. When the ego is in charge in the human psyche, we are unlikely to act in the common human good and instead favor the well-being of the in-group we feel we belong to over everyone else. Thus, there is a three-way reciprocal relationship between ego, concentrated wealth and power, and the climate crisis, as illustrated in Figure 3.1.

Unfortunately, those who are unconscious of their egos may be especially willing to further degrade the quality of our ecosystems in order to achieve what they consider to be their own success. Our challenge today is that if everyone's ego took charge and put self-interest first, the result would be more than 8.1 billion people competing endlessly and greedily for resources, wealth, and power on a beautiful, bountiful, yet limited planet. The result would be the climate crisis, along with other related global survival threats, such as xenophobia, preparations for war, and pandemic.

In this text, we are not suggesting that we wage war on the ego and on other left-hemispheric characteristics, but rather that we rebalance our consciousness and temper our ego by empowering right-hemispheric thinking as well. We know that the left hemisphere (like the beige wolf in the story in our preface) has many important roles in human functioning, including logic, scientific reasoning, and language processing. We also have evidence that the right hemisphere tends to take a more nuanced and wholistic view of language and the world and has a more direct and sensual relationship with the physical world of nature and our own bodies (McGilchrist, 2009).

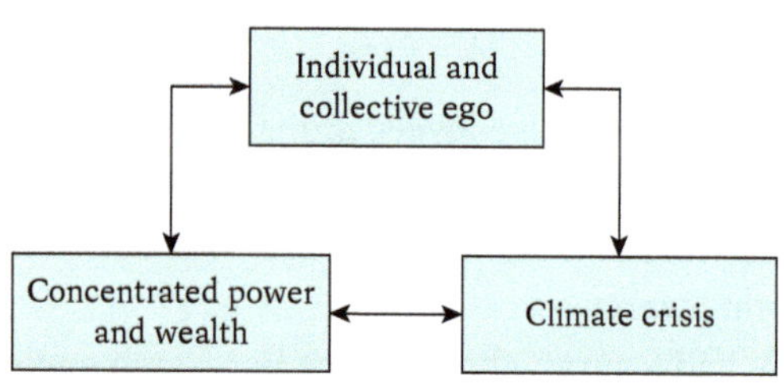

FIGURE 3.1 The Reciprocal Interrelationships Among Concentrated Wealth and Power, Ego, and Our Climate Crisis

In other words, the right hemisphere can help us temper the ego's focus on self-interest and help us see the wisdom in an *enlightened self-interest* that understands that the well-being of humanity is ultimately in everyone's self-interest.

In a world threatened by the climate crisis and other global survival threats, such an enlightened self-interest is about the importance of the collective, highest good. And, in fact, it may well be that humanity is already in the process of evolving toward a new tempered ego that recognizes the importance of such an enlightened self-interest, which puts the well-being of the collective whole of humanity, living things, and ecosystems over individual or small-group success.

Enlightened self-interest seems to be rooted in a combination of both genetic and learned origins. We know that our ancestors survived because they were willing and able to cooperate with each other in their hunter-gatherer groups (Vince, 2020). The potential thus seems to exist for contemporary humans to develop a new enlightened self-interest that is more global in scope. We have a tool to use to help transform the climate of our own consciousness, which is our ability to view ourselves and the world with reverent awareness.

When we assist people with their ego work, we want to help them become conscious of both their individual and collective egos. I am being influenced by my individual ego when I identify with any *personal* belief, including religiosity, political views, ideas about vaccinations, or convictions about climate change. In contrast, *collective ego* is an identification held by an entire group of people, from a family to a nation in size. For example, an entire caste of people in a particular nation might believe that they are either inferior or superior to those in other castes or socioeconomic groups.

We know that most human activity involves activity on both sides of the brain. However, egoic thinking appears to be especially associated with left-hemispheric functioning in the human brain. In his summary of the scientific literature, McGilchrist (2009) summarizes *left-hemispheric functioning* as being particularly "important for being able to get things from the world for one's own purposes, involves isolation of one thing from the next, and isolation of the living being perceived as subjective, from the world, perceived as objective. The drive here is manipulation and its ruling value is utility" (pp. 127–128). Thus, the egoic drive for self-interest, individuality, and identification with belief is congruent with left-hemispheric functioning.

In contrast, McGilchrist summarizes that *right-hemispheric functioning* moves us toward:

> the sense of the connectedness of things, before reflection isolates them, and therefore towards engagement with the world, towards a relationship with "betweenness" with whatever lies outside the self. With the growth of the frontal lobes, this tendency was enhanced by the possibility of empathy, the seat of which is in the right frontal expansion in social primates, including humans (pp. 127–128).

McGilchrist entitled his text *The Master and the Emissary* because he found that a dominant left hemisphere tends to silence our right-hemispheric tendencies, whereas the right hemisphere, when dominant, tends to take a more "both-and" approach to leadership and shares power with the left hemisphere. In other words, the left hemisphere is more controlling than the right and abuses any power it has. We can see evidence that this pattern of left-hemispheric dominance has been strengthening over thousands of years of Western cultures and has therefore become increasingly influential in a world increasingly controlled by Western culture.

These dominant left-hemispheric tendencies seem to create patterns of human thinking and behaving that are especially associated with our climate crisis. We can examine some of the characteristics of left-hemispheric dominance that McGilchrist's research identified. In Table 3.1, these are are listed in column 1, and all may well contribute to our climate crisis. In column 2, we list the characteristics of a more balanced consciousness in which both left- and right-hemispheric functions are felt and expressed.

TABLE 3.1 Human Consciousness and Ego-Hemispheric Imbalances

Characteristics of Consciousness When Dominated by Egoic Left-Hemispheric Functioning	Characteristics of a More Inclusive and Healthy Consciousness (Equally Influenced by Both Hemispheres)
A sense of both impotence (inability to help the world) and moral superiority over others	Connection with other people and empowerment to make a difference
Emphasis on each person's or group's own immediate self-interest	Emphasis on the well-being of all people, all living things, and ecosystems
Narrow view favored over broader picture	Ability to see both narrow and broad views
Emphasis on specialization in science and technology	Drawing from all disciplines in knowing the world and creating new technologies
Focus on detail and the downplaying or ignoring of uncertainty	Exploration of known and possible previously unknown consequences before acting
Loss of appreciation of context	Recognition of the context of human consciousness, including our culture, conditioning, and identities
Replacement of common sense, intuition, and imagination with only rationality and logic.	Honoring all ways of knowing and using them to check and balance each other
Knowledge valued over wisdom	Use of both wisdom and knowledge in decision-making
Expertise and experience replaced by expert knowledge	Value of both experience and expertise as well as knowledge when making plans
Real work replaced by documentation, planning, paperwork, management, and other bureaucratic procedure	Value and allowance of ample time for our "real work" of professional helping, which is our practice in service to the world
Emphasis on how much you can do and on how quickly and precisely you can do it	Recognition of the extent to which we contribute to collective well-being
Widespread competition and uniformity that is associated with fear and resentment	Protection of democratic and inclusive processes in which all voices are heard
Skilled altruistic roles of professional helpers being both highly regulated and devalued	Appreciation and support for professional helpers while continued maintenance of accountability for and excellence in our actions

Source: Derezotes, 2023.

How are the characteristics in column 1 associated with the climate crisis? We can go through the list briefly to explain. The combination of impotence (feeling helpless to change the world) and moral superiority (believing that others are more to blame than myself) is arguably associated with human inaction and lack of cooperation with regard to any crisis. Obviously, a belief in my own powerlessness can lead to inaction on my part, and the sense of superiority might help me avoid any feeling of responsibility to change my inaction. The characteristic of immediate self-interest might also serve to make me avoid taking any responsibility for the climate crisis. The narrow view is a short-term one that reinforces my own immediate self-interest.

The current emphasis on specialization in science and technology (knowing more about less) separates us into monodisciplinary boxes rather than multidisciplinary dialogues. As we discussed, the climate crisis is a global phenomenon with elements within the purview of many disciplines of study. With a diversity of many viewpoints and skills, a multidisciplinary team is arguably much more likely to transcend a focus on unnecessary details and also identify both the obvious and the more hidden consequences of their decisions. Along these lines, a diverse team also is positioned to notice the context of climate crisis issue in complex and unique local and more global contexts. Related to this, a fundamental belief in inclusive practice leads to the use of all ways of knowing when dealing with micro-, mezzo-, macro-, and eco-level challenges.

Although knowledge is vital to our work with the climate crisis, the world also needs *wisdom*, which is the ability to apply our knowledge, especially to new and complex situations. Such wisdom can be fed by experience and expertise. Helping professionals need more support in working with the climate crisis–related issues most of our clients face today rather than with just more restrictions from private insurers on what our work should be about and increased paperwork requirements to justify what we do. Similarly, there is little reason to think that in our time of climate crisis, helping professionals should be under unnecessary and excessive pressure to demonstrate how much we can do and on how quickly and precisely we can do it. There is, however, ample reason to think that most people would benefit from living and working in inclusive institutions and communities that value our professional helpers and have democratic processes that are protected and in which all voices are heard.

Grandiosity, Impotence, and Moral Superiority in Human Consciousness

We can think of grandiosity as one of the hallmarks of egoic thinking and behavior. Helping professionals will observe grandiosity in most of our clients and inevitably in ourselves as well. When any of us has an unexamined ego, they become more susceptible to grandiosity. We can think of *grandiosity* as an attitude of excessive self-importance and specialness that can be associated with such attributes as lack of empathy, superiority, boastfulness, competitiveness, rule-breaking, and defensiveness (Raypole, 2021). Although these traits in moderation may be useful and appropriate in certain situations, a long-term pattern of grandiosity, as helping professionals know, can lead to self-destructive behavior and feelings of isolation and aloneness.

It may be helpful to further explore the nature of our grandiosity by using continuums again, which we introduced as part of inclusive practice in Chapter 1. Figure 3.2 illustrates how our grandiosity can be understood as a characteristic of ego that can show up in either (or both) of two seemingly opposite ways: shame or narcissism. And, in either case, shame and narcissism can be understood as the two ends on a

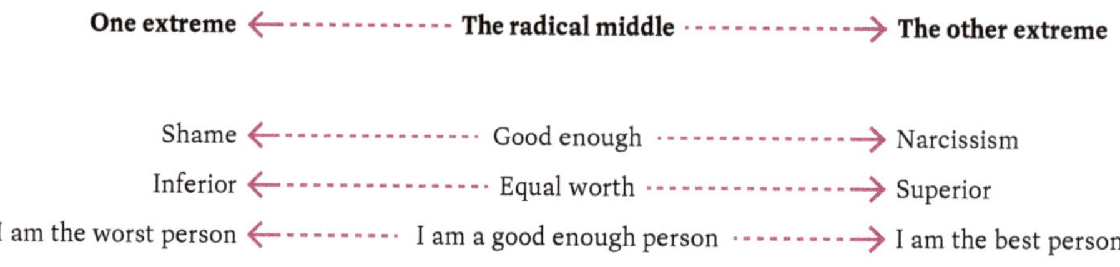

FIGURE 3.2 Grandiosity: Two Sides of the Same Coin

continuum of grandiosity—I am either the very worst or best person. When I shame myself, I judge myself as inferior to others in my family, community, or culture. When I take a narcissistic position, I compare myself with others and see myself as superior to them. As illustrated, if we take the radical middle position, we choose to see ourselves and others as "good enough." We will build on this idea in the text and theorize that the person who takes the middle position of being good enough is most likely to engage in effective activism.

We can observe an example of grandiosity at the macro level in the sociology of the climate crisis in the United States. The Science Resources 3.1 box presents a summary of findings from polling research on our attitudes toward the climate crisis. We can see that although most people in the United States are concerned about the climate crisis, most of us are also pessimistic about our chances of succeeding in turning things around. Many people report in these polls that their pessimism is based on their belief that humans are averse to changing our behavior. We can theorize that this apparent contradiction about expressed concern and inactivity has something to do with our egoic nature. It is likely that many of us who lament the inactivity of others regarding the climate crisis use this fact to justify our own inactivity. In other words, we humans have a tendency to project the traits we do not want to see in ourselves—in this case, our inactivity—onto others around us. Again, "if you spot it, you've probably got it."

SCIENCE RESOURCES 3.1 What do we believe, and how do we behave about the climate crisis?

IMG 3.2

- Recent polling in the United States suggests that 76% of those polled report that they are very or somewhat concerned about the climate crisis.
- About 66% of us are pessimistic about humanity's chances of responding effectively to the climate crisis, primarily because we believe that most people are not willing to change.
- A total of 49% say that they have experienced extreme weather.
- Regarding the other 51% who report that they have not experienced extreme weather, only 24% of them say they are very concerned about climate crisis.
- A total of 68% of those polled believe weather extremes will become more frequent.

- A total of 39% experience extreme weather events as harming their everyday life.
- Pollsters indicated that these percentages varied across geographical location and political affiliation.

Source: Weise, 2023.

A majority of people in the United States currently report that the climate crisis is a lower priority for them in comparison with such issues as the economy and health care costs. However, 65% of Americans are worried about global warming.

In fact, about two-thirds of people in the United States report that they rarely or never discuss the climate crisis with family and friends. According to experts, we often choose to deny the climate crisis to protect ourselves from difficult emotions, and constant climate-related anxiety and stress do not seem to motivate positive action. However, others report that their anger can motivate their environmental activism.

Younger people are more concerned. Researchers have found that of 10,000 young adults worldwide, about 75% of folx between ages 18 and 25 find the future frightening (Randall, 2023).

Although most of us *are* in fact probably resistant to change in a world that is constantly changing, we could theorize that our willingness to blame our own lack of activism on the resistance to change in other people may be a psychological—and probably largely unconscious—defense, which may be an attempt to shift the guilt and shame we feel about our own sense of helplessness and inactivity.

As we discussed earlier in this chapter, McGilchrist (2009) observed that many of us approach the world with an attitude of impotence, which is a left-hemispheric attitude that we have little, if any, ability to make a difference. This attitude seems to be common in the consciousness of many people in our country today. In fact, this attitude of impotence may be rooted at least in part in the grandiosity of thinking that I am *suffering more* than anyone else on the planet. The thinking may be that since I am so disempowered, unimportant, abandoned, or alone, it is therefore not my fault that there is so little I can do to make a difference.

We can also see grandiosity at the micro and mezzo levels in our individual practice with our clients. Many people approach their daily lives through a lens of grandiosity, which means that we often approach the world not only from a position of impotence, but also from a position of moral superiority. These two traits, seemingly opposed, actually work together. Remember that the ego acts primarily according to self-interest. The climate crisis, as we have seen, is a complex and potentially overwhelming global threat to our well-being and existence that challenges the ego's desire to live a predictable and safe life. Moral superiority can help me compensate for my guilt about inactivity and can also be another way to defend against the climate crisis threat.

For example, if I minimize or deny that there is a climate crisis, then I can feel superior to other people whom I believe are being duped or controlled by those "radical climate extremists." I can also focus my activities on some other moral or cultural issue that does not require me to make any uncomfortable life changes but serves to help me feel morally superior to those who disagree with me.

Perhaps the most important tool we have in dealing with the ego-related grandiosity inherent in impotence and moral superiority is, again, our own reverent awareness. A climate change in our consciousness begins with greater awareness and acceptance of who we are, in the here and now. Helping professionals can assist individuals, institutions, and communities develop their consciousness through counseling, psychoeducation, and community events. In Section 2, we will review how dialogue can be used in community events to transform consciousness and enhance climate change in humansystems.

Self-Interest, Individuality, and Collectivism in Human Consciousness

We theorized earlier in our chapter that left-hemispheric, including egoic, dominance in our consciousness is linked to self-interest and individuality. According to Cheng et al. (2020):

> Individualism is the belief that individuals are separate beings with personal differences that make him/her unique from the group. Internal processes guide individualists' behaviors, with the goal of reaching self-satisfaction and one's full potential. Collectivism is defined as a social pattern that consists of individuals who are closely interconnected in a group. Collectivists gain their values and social norms from the group. They are more likely than individualists to identify with the intergroup pattern. (p. 1)

The self-interest and individuality associated with the ego and left hemisphere probably exists to some degree in all human consciousness. However, we can especially see a tendency for individuality to be expressed today in the more Western cultures such as Western Europe, Australia, New Zealand, South Africa, and North America. In such cultures, traits like individual uniqueness, competition, fame, self-expression, and self-development are emphasized. We can also see that the limitations of individuality may include the lack of concern for such things as collective well-being, empathy for others, mutual support, connection, and cooperation. In contrast, more collectivist cultures, which emphasize concern for group well-being over that of individuals, may be commonly found, for example, in cultures in Southern Asia, including China, Korea, and Japan. In these cultures, the limitations of collectivism might include an inability to have personal boundaries, a lack of self-esteem, and a lack of self-development (Wolf, 2023).

Fortunately, there is a middle ground between extreme individualism and extreme collectivism. In this radical middle position, the person values elements of both individualism and collectivism. In Figure 3.3, we illustrate this idea by drawing a continuum that goes from individualism on one end to collectivism on the other, with the idea of balance located in the radical-middle position of the line.

What might this balance between individualism and collectivism look like? It might look different in different contexts. In a family, for example, who values both extremes, every member has a voice, but the parents still make the final decisions about major finances, major purchases, and most house rules. In an institution, for example, in a community college, all staff, faculty, and administrators might have the right to debate and vote on many major issues without undergoing institutional punishment or other negative

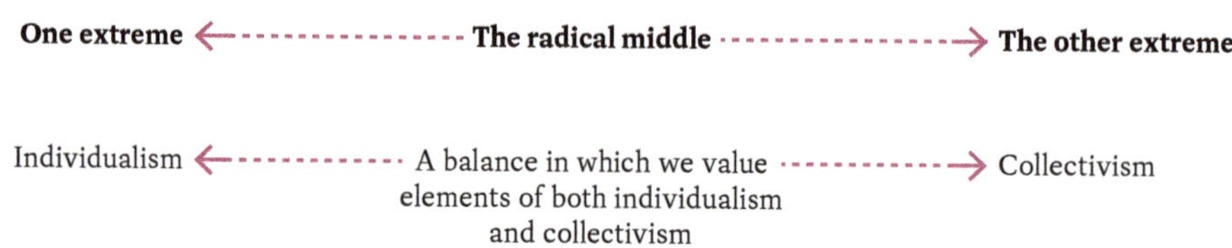

FIGURE 3.3 Continuum of Individualism to Collectivism

consequences. However, administrators at the college might need to still make decisions in various everyday and long-term campus functions.

Most human beings seem to need at least some psychological defenses to cope with an unpredictable and challenging world that is constantly changing. Examples of common coping defenses might include denial, projection, intellectualization, and rationalization. The healthiest defense is one that is conscious and flexible, one that enables me to choose when and how I can best cope and defend myself in each unique situation.

It seems obvious how self-interest, individuality, denial, and grandiosity can all be associated with our climate crisis. All three of these interrelated traits can both rationalize my misbehaviors toward the environment and also serve as psychological defenses that protect me from the painful reality of the climate crisis. An example of self-interest might sound like, "I am too busy to sort out my garbage and take items to the recycling plant." Individuality might sound like, "Those environmentalists all want to tell us all what to do. I'm not going to let them boss me around!" And grandiosity and denial might sound like, "I know that the air quality is bad, but this idea of climate crisis is just a hoax perpetuated by the elites. Let *them* cut down on their driving. I have a lot of important things to do right now and can't be bothered."

Helping professionals want to assist our clients in examining the extent to which their coping styles are habitual and unconscious. We can help our clients become more conscious, for example, of their grandiosity, individuality, denial, and self-interest. A sense of moral superiority, for example, as we have discussed, is a common form of grandiosity we use today to protect ourselves and cope with the world, but like all coping styles, it can sometimes be unhelpful (such as when we become furious after being stopped by the highway patrol for speeding). Similarly, a habitual pattern of self-interest might sometimes serve us well but will also be self-defeating in particular situations. I like to suggest to my students and clients that they consider developing an attitude of enlightened self-interest in which they consciously find the radical middle, balancing personal and collective interests in each unique life situation.

Living in This Age of Crisis

Perhaps we all have a bit of crisis fatigue. When I streamed the news this morning, the commentators were talking about "the crisis in the classroom" in our state. I noticed my first emotional response was to push away the news since my life feels already full enough of crisis. Often the first response that we have to any news is ego-based. This one was.

After becoming curious about my reaction, I did some research and discovered that the ancient Greek root word for *crisis* is *krinein*, which meant to "separate, decide, judge, explain" (Online Etymology Dictionary, 2024). In other words, some of our ancestors thought of a crisis as a situation that challenges us to see what is actually happening and to make a decision about what to do in response. I don't think my ego likes the kinds of crisis we have today because they all call for an honest self-examination of the world and my own consciousness and for me to take some kind of action rooted in that understanding. An honest appraisal of my reactions to the climate crisis is emotionally difficult. My ego is self-absorbed and does not especially want to think about other people, other living things, and the ecosystems that support life. And my ego has too much grandiosity to take a look at how I contribute every day toward making the climate crisis even worse.

In fact, it seems there are perpetual multiple crises in this current era in which we all live today that demand that I separate, decide, judge, and explain. It might help me to cope better if I seek to *explain* the roots of these challenges to myself and others, for example, and find any interconnections. Then, it may be important to *separate* and order the crises in their order of importance and *decide* on which to focus on first. I also possibly want to honestly *judge* how effective my responses are.

All these tasks may require painful emotional and cognitive work, and my initial egoic reaction might be to minimize or deny the crises of which I am aware. Although we all need a break sometimes from difficult news, we also pay a price when we habitually avoid difficult news, which might include anxiety and depression. We can understand both anxiety and depression and even despair as related defenses against stress, in which we enter our own head and ignore how the rest of our body may feel. In other words, we numb ourselves. This tendency to numb ourselves in reaction to what is going on in the world is widespread and serious and can be linked with both anxiety and depression. Depression has been considered a global health problem for many years by the World Health Organization (2023) and associated in part with the climate crisis.

Professional helpers can help our clients and communities reframe depression and anxiety and despair as understandable human reactions to—often at least in part—the challenging world we live in today. In other words, we can help people become more conscious of our own symptoms and where they come from. And, as we examined in Chapter 2, as we allow ourselves to experience and accept our emotional reactions and feelings related to the climate crisis, we can start to heal and become ready to join the many other people who have become climate activists.

Throughout the text, we emphasize the role consciousness, or reverent awareness, plays in our healing and transformational work. Such reverent awareness is usually emotionally difficult because we are opening up our hearts and minds to seeing the world the way it actually is. The ego, as Tolle (2006) points out (and as quoted at the top of this chapter), "The moment you become aware of the ego in you, it is strictly speaking no longer the ego, but just an old, conditioned mind-pattern. Ego implies unawareness. Awareness and ego cannot coexist" (p. 1).

Keeping My Ego Tempered in My Activism

Those who deny that there is a climate crisis are not the only people with egoic issues. Some of the most committed climate activists I know who strive to educate the public and effect social policy change have also surrendered to their ego. What do I mean by that?

Remember that the ego always thinks it is right and carries a feeling of moral superiority over others. This can certainly happen to activists, regardless of how intelligent, informed, and committed we are. Some common symptoms that may suggest that my ego is taking over my good work in dealing with the climate crisis include the following:

1. I find myself getting quickly angry and defensive with anyone who disagrees with me about the climate crisis.
2. I start comparing myself with other people, especially those who disagree with me, and feel either superior or inferior to them, especially when under stress.
3. I find myself arguing with other activists about the right strategies we should use to save the planet and stop listening to other ideas.
4. It becomes more important to win arguments or conflicts about the climate crisis than to cooperate with others who disagree with me for the ultimate highest good of all.

Generally, when my ego gets involved in my activism, I become less effective. Other people can sense the egoic nature of my thinking and behavior and tend to react to my ego rather than to what I am saying and doing. My responsibility as a leader and activist is to strive to stay conscious of my ego and left-hemispheric functioning.

Ego Work, Across the Developmental Dimensions

How can we build on these insights into our ego-hemispheric imbalances to change the climate of our human consciousness and effectively address the climate crisis?

We can describe examples of consciousness work in the human dimensions of development, described in the inclusive practice term *ecobiopsychosocialspiritual*. These dimensions include the physical, emotional, social, cognitive, spiritual, and ecological. In the following section, we describe some interventions that can help our clients increase their consciousness while working in each of the dimensions.

Physical Development

The professional helper can help the client become more aware of how their body feels when they come into contact with climate extremes. We can ask for the client to become more conscious of what is it like, for example, to live in extreme heat for days and even weeks at a time, as increasing numbers of people in the United States and other nations now do. Or we can ask our clients how the stress from threats of flood and wildfires feels in their bodies.

Some clients may be relatively out of touch with their body. They may therefore need to spend time each day reflecting on what their bodies are telling them. For example, maybe one client seems to have headaches every Sunday evening, which they learn has something to do with the dread they feel about going back to work on Monday morning. Or consider a professional athlete, who never learned to rest when their body is tired, who finally discovers the benefits of listening to what their body needs.

Emotional Development

We could say that emotion often puts us into motion. Whether we feel sad, mad, or glad; scared, excited, or disgusted, emotions can motivate us into new ways of thinking and acting. The professional helper can assist clients in becoming more conscious of their emotions and learning new ways to express their emotions effectively. And since we understand *empowerment* as the ability of a person to be conscious of and express their emotions and thoughts, we could thus say such emotional work is self-empowerment work.

I have some favorite interventions that I use with clients. For example, I might ask them to keep a journal about the emotions they experience each day. Sometimes I suggest that families put a white board or bulletin board up in the kitchen on which family members can indicate the strongest emotions they are experiencing. I also like to do check-in exercises before a class or training, in which I ask people to briefly share how they feel, as well as "check-out" exercises at the end of the group time, during which people can briefly share their emotions again. When I ask them, many of my students and clients have told me that they have had dreams at night about the climate crisis that they may need to share.

Cognitive Development

The professional helper can help the client examine their past experiences that may still influence the client's egoic functions today. We can ask the client to explore, for example, how they learned to feel superior (narcissistic) or inferior (shame) as a child growing up. Or there can be discussions about how grandiosity may have been modeled by key figures in their life and how the client reacted to that modeling.

The emphasis in this work is on developing consciousness about what happened and how past experiences still influence the present moment, especially in how the client learned to interact in their relationships with other people, other living things, and ecosystems.

Social Development

One approach to social development is to help the client develop greater consciousness about how they currently think and act in relationship to morality.

For example, journalist and cultural and political commentator David Brooks (2023) has called for a moral education for our youth. In such an education, students are not told what to believe and how to act but rather are asked to wrestle with the following challenging questions:

- What is the strongest value you have in your life?
- To what people, living things, and ecosystems do you feel a responsibility?
- How would you best live a meaningful life?
- How does a moral human live today?
- By what principles should we make decisions about technology, living things, and our environment?

Spiritual Development

As we discussed earlier, spirituality is not necessarily aligned with a religion, and increasing numbers of people today report that they are spiritual nut not religious. What does this mean? It means different things to different people, because spirituality is an individual element of human development, in contrast with how religiosity is a social element in life through which people may share rituals, doctrines, and beliefs. In our lifelong spiritual development, people can learn to recognize the connection we call have with other humans, other living things, our planetary ecosystems, and the rest of the immense universe in which we live.

Much of what we wrote about in this chapter can be viewed as related to the spiritual dimension. In our spiritual development, we find connection, meaning, and purpose in life as we extend and deepen our consciousness, both inward into who we are and outward into the world. There is an element of surrender required in the spiritual journey, as we become conscious of our egoic nature and learn to let it go. The idea of death and rebirth may be helpful, as we learn to let go of identification with the things that are not truly who we are.

Many people find nature to be a spiritual teacher. In a number of wisdom traditions, great spiritual teachers often entered natural settings, such as mountains or deserts, to find inspiration and do their spiritual work. For example, Tolle (2006) writes about appreciating the ethereal nature of a flower. The origins of the word *ethereal* refer to the highest parts of the atmosphere and to a sense of the light and airy. The beauty of lowers can be inspirational and hint at the spiritual qualities of living things and sacred landscapes in our world.

Ecological Development

We can think of ecological development as having to do with my relationship with nature. What we refer to by the word *nature* is spaces in which we humans can have some interactions with other living things and the ecosystems that support all life.

I often give my clients assignments to go out in nature, which can involve anything from urban gardening to river rafting or solo camping. I will usually ask them to take a journal and write down their thoughts, emotions, and behaviors. I might ask them, "How do you feel when you are outside?" or "What do you think about when you are alone with yourself for a day outside?" Some of my clients choose to go on vision quests, during which they find a place to camp for a night or two and reflect and journal.

Frequently, my clients report to me how their consciousness changes when they are outdoors. They may report that they "feel more like they are in their body," that they are calmer, or that they "cleared their head." I believe that people often deepen their consciousness; they become more aware of themselves and the world and even more accepting. Many people report having a spiritual experience when outdoors, particularly when they visit a natural place that is sacred (i.e., revered) to them.

An ecologically mature individual is in touch with their biophilia. *Biophilia*, which literally means "love of life," refers to a biological need for humans to affiliate with living things (Wilson, 1984). We probably could also create a theory of *ecophilia*, in which we need to also affiliate with natural landscapes, such as cloud formations, mountains, rolling hills, beaches, and steppes.

Finally, evolutionary biologists have found that copying the behaviors of others can be an extremely effective way to influence new behaviors and is actually much more powerful than innovation in transforming human behavior (Vince, 2020). In other words, we can teach people through modeling how to live more sustainably in a world threatened by our climate crisis. Helping professionals can model new ways of thinking and acting using all of the ecobiopsychosocialspiritual dimensions of development, as appropriate.

QUESTIONS FOR REFLECTION

1. How have you experienced your own ego? Please describe in your own words how it feels inside of you, how it influences you, and what kind of relationship you have with it today.
2. Do you think your left-hemispheric functions dominate your right hemisphere? Please explain.
3. Some philosophers may ask whether all private goals are unhealthy. What do you think? Can there be such a thing as enlightened self-interest, as we discuss in this chapter?
4. Is it true, as Tolle suggests, that when we are conscious of our own ego, it cannot exist? Please explain your own experience and thoughts about this idea.

REFERENCES

Brooks, D. (2023). How America got mean. *The Atlantic*, 68–76.

Cheng, A. W., Rizkallah, S., & Narizhnaya, M. (2020). Individualism vs. collectivism. In B. J. Carducci, C. S. Nave, J. S. Mio, & R. E. Riggio (Eds.), *The Wiley encyclopedia of personality and individual differences: Models and theories.* Wiley Online Library, September 18, 2020. https://doi.org/10.1002/9781118970843.ch313

Derezotes, D. S. (2023). *Inclusive social work practice: A new vision of community practice.* Cognella.

Goodreads. (n.d.). *Consciousness quotes.* https://www.goodreads.com/quotes/tag/consciousness

Maté, G. (n.d.). *Gabor Maté: 40 powerful and life-changing lessons to learn from Gabor Maté.* Purposefairy.org. https://www.purposefairy.com/97666/gabor-mate-life-changing-lessons/

McGilchrist, I. (2009). *The master and his emissary: The divided brain and the making of the Western world.* Yale University Press.

Online Etymology Dictionary. (2024). *Crisis.* https://www.etymonline.com/search?q=crisis

Randall, R. (2023). Spirituality, global warming, and grief: How clergy can help tackle climate anxiety. *Mother Jones*, June 18, 2023. https://www.motherjones.com/environment/2023/06/spirituality-global-warming-and-grief-how-clergy-can-help-tackle-climate-anxiety/

Raypole, C. (2021). *What is grandiosity?* PsychCentral, April 28, 2021. https://psychcentral.com/blog/grandiosity-and-delusion-grandeur

Tolle, E. (2006). *A new earth: Awakening to your life's purpose.* Plume.

Vince, G. (2020). *Transcendence: How humans evolved through fire, language, beauty, and time.* Basic Books.

Weise, E. (2023). USA TODAY, Ipsos poll: 20% of Americans fear climate change could force them to move. *USA Today*, September 6, 2023. https://www.usatoday.com/story/news/nation/2023/09/06/climate-change-divides-america-usa-today-ipsos-poll-data-shows/70533243007/

Wilson, E. O. (1984). *Biophilia.* Harvard University Press.

Wolf, M. (2023). *The crisis of democratic capitalism.* Penguin.

World Health Organization. (2023). *Depression.* https://www.who.int/health-topics/depression#tab=tab_1

Credits

IMG 4.1 & 4.2

To Be or Not to Be

Eros and Thanatos

To be, or not to be: that is the question.
—WILLIAM SHAKESPEARE (1599–1601 EST.)

Men have brought their powers of subduing the forces of nature to such a pitch that by using them they could now very likely exterminate one another to the last man. They know this—hence arises a great part of their current unrest, their dejection, their mood of apprehension. And now it may be expected that the other of the two heavenly forces, eternal Eros, will put forth his strength so as to maintain himself alongside his equally immortal adversary.
—SIGMUND FREUD (1930, P. 144)

Most of us have at least one time in our life when we are in deep suffering, we think about life and death, and we consider the prospects of suicide. My time was in my 21st year. It was the fall of 1971, the days were getting shorter, cold rain was already falling, and the Beatles were no longer a band. Lennon's song "Imagine" had been released, and McCartney had formed his new group, Wings. The U.S. military was still in Vietnam, and we all knew someone who had died over there. While living with my college friends in our commune near campus, I participated in various experiments with relationships, majors, hallucinogens, music, and life meanings.

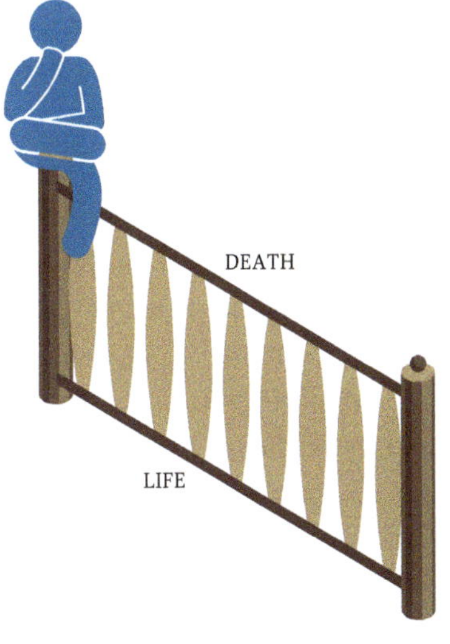

IMG 4.3

I had become anxious and had not discovered yet that these symptoms were an opportunity to start healing my own past trauma. I did not know yet that I was sitting on the fence between life and death, but I did know that I was done feeling so unhappy.

And I would soon find out that although life could be worth living, there was a part of me that actually needed to die.

The Archetypes of Eros and Thanatos

To be or not to be, that *is* the question and the greatest challenge facing humanity today: Will we decide to survive and prosper, or will we decide to destroy our world and ourselves with our ever-growing technologies?

In this chapter, we will explore these two choices we have in front of us. We will represent these two directions with two archetypal figures from Greek mythology, *Eros* and *Thanatos*. We will also use the Jungian concepts of *shadow* and *light* to examine the suppression and consciousness of Eros and Thanatos (Taylor, 2021).

We use the term *archetype* in this text to describe patterns of human experience, emotions, thoughts, and behavior. The history of archetypes can be traced back to the beliefs that our ancient ancestors had in gods and goddesses who personified common elements they experienced, including the seasons, harvest, justice, sky, seas and mountains. These personifications were found to be useful in explaining, and dealing with life events. The philosopher Plato created the term *archetype*, which was derived from ancient Greek words that meant "original pattern".

The psychologist Carl Jung (see https://www.jungpage.org) saw how these ancient archetypes are similar to archetypes that seem to still exist in our lives today, and how we can work with these often unconscious symbols to find greater understanding, purpose and meaning. Professional helpers can sometimes use archetypal psychology as another tool in helping people

Eros is the ancient Greek god of love and life, and Thanatos is the ancient Greek god (also called the *personification*) of death and darkness. Eros was the son of Aphrodite, goddess of love and beauty, and Ares, god of war. Thanatos was the son of Nyx, goddess of the night, and Erebus, god of darkness.

Eros and Thanatos in Light and Shadow

During his career, Sigmund Freud (1930) developed the theory that humans make decisions based upon two opposing "life instincts": the life drive and the death drive. Whereas the life drive is about life, survival, procreation, and cooperation, the death drive was about death, aggression, and release from suffering. For Freud, decisions based on the life drive are driven by love and a desire for pleasure, happiness, affection, and community. In contrast decisions based upon the death drive are driven by anger and impulsivity and perhaps even a desire for homicidal or suicidal behavior. Freud identified as Jewish and was living in Germany when Hitler came to power; there were terrifying examples of Thanatos in his world, just as there are such examples today again in our world.

In *Life Against Death: The Psychodynamic Meaning of History*, Norman O. Brown (1959) built on the psychoanalytic theories of Freud. Brown followed the lead of other authors in using the terms *Eros* and *Thanatos* to describe the life and death instincts. He suggested that humans may be able to heal the divide between Eros and Thanatos in a kind of dialectic in which both instincts are healed and resolved into a coherent synthesis. For Brown (1959), the life, or Eros, instinct "demands a union with others and with the world around us" (p. 307), whereas the death, or Thanatos, instinct can be "affirmed in a body that is willing to die" (p. 308). Brown finishes his text with a call for humanity to become "friends" again with Eros.

Brown continued what he called his "voyage" into psychoanalysis in *Love's Body* (1966). In many ways echoing McGilchrist's (2009) neuroscience that we discussed in earlier chapters, Brown called for a healing of all the dualities in human consciousness, a "reconciliation" of body, mind, and spirit, or what McGilchrist might call a rebalancing of the right and left hemispheres. According to Brown:

> Personality is persona, a mask. The world is a stage, the self is a theatrical creation. ... Freud came to give the show away; the outcome of psychoanalysis is not "ego psychology" but the doctrine of "annata" or no-self: the ego is a "me-fabrication" (*ahamkara*), a piece of illusion (*Maya*) which disintegrates at the moment of illusion (p. 90).

With these words, Brown links the rising of consciousness that can come from Western psychoanalysis with the discoveries of the spiritual psychologies of other cultures in the East and South. In Chapter 15, the last chapter of this text, we will further explore the relationship between consciousness and the climate crisis.

Perhaps the life and death instincts do not have to be dualistic adversaries in our consciousness. According to Kli (2018), a nondualistic interpretation of Eros and Thanatos is possible. If we interpret that both drives are manifestations of the same "libido" drive, the opposing needs for pleasure and self-fulfillment and the need for self-preservation and self-maintenance can all be fulfilled. A "return" of Eros is called for, writes Kli, in which humanity rediscovers and embraces love, pleasure, wholeness, and vitality. Perhaps, as we will suggest later in our chapter, we can heal the divide, and both Eros and Thanatos can indeed serve the individual and collective consciousness of humanity and, together, help us deal effectively with the climatc crisis.

Building on this very brief introduction to a complex subject, we will use the terms *Eros* and *Thanatos* to describe the two seeming opposing forces observed inside of us today. We will examine the growing dominance of Thanatos in this age of anger and, like Freud, Brown, and Kli suggest, search for ways to also rediscover and express Eros more in our world today. We will view Eros in this text as including not only romantic or sexual love, but also *storge*, or familial love; *philia*, or brotherly (friendship) love; and *agape*, a giving and caring love for others and the world. We will look at both Eros and Thanatos as potentially valuable "parts" of human consciousness that can be healed so that they are both become accessible to inform human survival and well-being.

In addition, as introduced earlier, we will examine what happens when we individually or collectively suppress Eros and Thanatos, thus making them unconscious and pushing them down into the "shadow" of our awareness. The shadow is the parts of ourselves that we do not want to have (see a detailed Jungian explanation of shadow and light at https://www.jungpage.org). We will also examine what happens when we shine the light of consciousness on Eros and Thanatos. Table 4.1 summarizes these four main ideas, which are examined in this chapter. We can see that when we push Eros or Thanatos into shadow, we express these instincts in more unhealthy ways, and when we shine the light of consciousness on Eros and Thanatos, we express them in healthier ways.

TABLE 4.1 Eros and Thanatos in Shadow and Light

	When in Shadow (Unconsciousness)	When in Light (Consciousness)
Eros (life)	Because I feel disconnected from myself and the universe, I suffer and may unconsciously seek ways to temporarily find pleasure and avoid pain, often through addictive behaviors that ultimately harm myself and the world.	I am conscious that I am connected with everything, which can motivate me to express loving kindness toward myself, other living things, and the world.
Thanatos (death)	Because I deny the temporary nature of everything, I may unconsciously identify with what I think I can control, such as beliefs, wealth, power, and fame, and ultimately cause harm to myself and the world.	I am conscious of the reality of death, which can motivate me to disidentify from those things that are not me and have gratitude for what I am and what the world is.

As I write these words this afternoon, we are watching the war develop between Israel and Hamas in the Middle East. In reaction to the terror attacks in Southern Israel a week ago, the Israel Defense Forces are amassing troops and weapons on the border with Hamas, but most of us are uncertain what will happen next. It strikes me that these ideas about Eros and Thanatos may also be useful to helping professionals in dealing with all the global survival threats with which the world is challenged today, including the climate crisis, preparations for war, and mass-casualty terrorism.

Sitting on the Fence Between Eros and Thanatos

At the beginning of the chapter, we have an image of a human figure sitting on the fence. Imagine that on one side of the fence is Eros and on the other is Thanatos. We can think about the image as representing the plight of many people in our world today. Fearful of Thanatos, we often hold back from forming a relationship with death. Equally fearful of Eros, many of us hold back from engaging in life as well. And, as we all know, it is not good to sit on a fence for very long. Your butt may start to hurt, of course, and you'll experience various other symptoms, such as anxiety, depression, and even despair.

Why are we afraid of death? One of my favorite textbook authors, Irving Yalom (2008), has written extensively about death and human consciousness. He explains that a fear of death is quite common in human beings in our culture, in large part because death is an ending that takes us into the unknown. Often we may not even be aware of our fear until we have an "awakening experience" in which we deal with an ending in our own life, perhaps, for example, the loss of a relationship, vocation, or our health. Yalom shows how most psychological issues with which our clients come to us for help are actually related to our fear of death. However, we live in a culture that denies the reality of death, which makes it difficult for us to explore how we actually feel and think about death.

Yalom uses the metaphor of staring at the sun to teach the concept of facing and accepting the inevitability of death. He encourages his students to use ideas drawn from the world's great wisdom traditions to help clients become more actively alive, develop relationships with other people, and find greater meaning and joy in their lives. As we will further investigate in this chapter, we can indeed become friends with Thanatos, and actually learn to "die a little every day," by letting go of things that no longer serve us so that we can live our lives more fully.

Perhaps less obvious is why we are also fearful of life. We could say that in order to participate in love and life, as Yalom suggests, we also need to have an open heart and mind. However, an open heart is also one that can be more easily broken again. And perhaps we may feel me need to keep our minds closed as well so that others do not take advantage of us. Thus, at least part of our avoidance of life is our desire to avoid pain. In addition, our shame also may make us feel unworthy of joy. Most people I know carry a feeling of shame through their life, which, from a psychological perspective, is an egoic sense of inferiority, often coupled with an attempt to compensate for shame with a feeling of superiority as well.

From a more spiritual perspective, shame can also be seen as a reaction to the disconnection from the universe that we humans feel at birth, and often express throughout our lives in the harm we bring to ourselves, other beings, and our environment. Perhaps this is why so many people who have had near-death experiences report that they felt for the first time in their life a wonderful connection with everything. We humans may all long for reunion with the universe throughout our life.

Related to shame, the dominance of left-hemispheric functions that we looked at in earlier chapters may also contribute to fear of life. Eros drives us to reach out to others, cooperate with them, and love them. The left hemisphere, as we saw in the last chapter, can also be a source of immediate self-interest, the narrow view, a sense of impotence, moral superiority, and competition. Perhaps many of us today live our lives "in our left heads," focusing increasingly on screen time, interacting all day with digital algorithms, and fighting our own negative thoughts, which contribute to our current global pandemic of anxiety and related depression.

Many of my students and clients have also discussed their fears of success in life. They talk about fear of increased responsibilities, expectations, and attention that might come with success. They also frequently discuss their fear of failure and how they've avoided applying for a job or meeting new people because of their fears of failure and rejection.

What Are the Origins of Eros and Thanatos?

Are Thanatos and Eros actually human instincts that evolved over millions of years of past evolution? Or are they actually traits that are learned during our lifetime, primarily modeled by the adults we knew in our childhood? Perhaps the best answer, with regard to our relationship with other people, other living things, and our life-supporting ecosystems, is both.

Peter Frankopan, in his *The Earth Transformed: An Untold Story* (2023), describes the history of humankind's relationship with the natural environment. He writes that throughout the Holocene, an era in history comprising the past 11,700 years since the ending of the last major ice age, we human beings have struggled to settle into, modify, and control nature. For most of the time humans have lived on Earth, we did this to survive. For example, some ancient famers burned the landscape and built water channels in order to improve the environment for planting crops, and some ancient hunters at some point in history began domesticating certain animals such as cattle, sheep, horses, and camels so that they could be kept nearby for food or transportation.

As Frankopan summarizes:

> The Holocene is the perfect tag for the modern age—a time that is geologically and climatically distinct, but which also corresponds to the era of modern humans, an era we invariably fail to place within the bigger picture. Since its beginnings ... our species has expanded, colonized, reproduced, created, and dominated, but it has also destroyed, devastated, and exterminated. It has done so better than almost any other organism that has lived in the past 4.5 billion years. (2023, p. 61)

Past of the reason that many of us humans are still willing to colonize, pollute, degrade, and destroy what remains of the Earth's biomes and anthromes may thus be our tendency to continue with the attitudes and behaviors that made us a "successful" species. In other words, our very desire for mastery over nature that helped us survive now also threatens our extinction. We may pass these tendencies on to our children and grandchildren through both genetics (instinctual behavior) and modeling (learned behavior).

Although professional helpers may not be able to alter genetic transmission of attitudes and behaviors, we may be able to educate people about how our ancient tendencies to seek blind mastery over nature may no longer serves us.

Healing and Transformative Work With Eros and Thanatos

As we have described, Freud, Brown, Kli, and Yalom have all called on us to form a new relationship with life and death, with Eros and Thanatos. We can think of this new relationship as a process of healing in which humans become "whole" and become friends with these aspects of ourselves. This healing process may typically begin when I become conscious of the extent to which I sit on the fencepost between life and death, avoiding both Eros and Thanatos. We may at the same time also start to see how the fear of life and death affects my life and the those of the people around me.

Our Suppression of Eros: Pushing Eros Into Shadow

A key role of Eros in Greek mythology was to bring love to the world. When people were shot with the arrow of love from the bow of Eros, they were overcome with loving passion. Interestingly, although originally a strong and powerful young man in ancient Greek stories, Eros has become much smaller and diminished over time, until he has now become current Western culture's infant, Cupid. It may well be that we have unconsciously tried to diminish the potency of Eros in our consciousness. Unfortunately, when humans try to ignore and lessen an instinct as powerful as Eros, we strengthen the "shadow" side of Eros. When we shine light only on Eros as a cute little infant who flies around on Valentines Day, we miss the powerful realities of Eros that we have pushed into shadow.

Eros is a connecting force in the ancient Greek stories. As we said, we are looking at Eros today as the drive for all forms of love, including *eros*, *storge*, *philia*, and *agape*. Therefore, Eros can be viewed as our instinct to "fall in love" with our families, communities, other people, other living things, and the universe itself. Our ability to appreciate beauty not only helps propel us towards connection with the universe, but, as Vince (2020) puts it, "We are made human by the contemplation of beauty" (p. 147).

It seems difficult, if not impossible, to suppress a part of me without suppressing more of me than I intended. When I suppress Eros, the love of life, I typically suppress elements of my own body, mind, and spirit as well. I tend to lose touch with my senses and might not notice, for example, what it feels like to be walking outside in a more natural environment or how my body feels when the seasons change. My feeling of joy is diminished, so I may habitually seek out any of the many substitutes for joy increasingly available to us today, such as fast food, electronics, lonely digital relationships, drugs, and alcohol. I also might not notice anymore what it is like to live in heavy air pollution or what it does to me to live in a city with so much light pollution that I cannot see the Milky Way on a moonless night.

And when I suppress my openness and ability to experience and enjoy life, I am likely to seek substitutes for the joy of living that I have lost. In other words, I may seek distraction in addictive behaviors that give me a moment of excitement but, in the long run, become increasingly deeply unsatisfying. When we become successful according to our culture's standards, it may be difficult to stay "sober" from the addictive qualities of fame, fortune, and power. Many of our most successful people, unfortunately, seem to mysteriously adopt self-destructive addictions that can lead to their self-harm and early death.

The Suppression of Thanatos: Pushing Thanatos Into Shadow

The key roles of Thanatos in the Greek pantheon was to select the people whose lives were over and then gently carry the souls of the dead to the underworld. Death may be one of life's biggest teachers, because it can remind me of the temporary nature of everything in life and thus wake me up to the joy of living in the here and now. However, we live in a culture that tends to deny and minimize the reality of death. When I suppress Thanatos, I deny the inevitability of death and lose the benefits in life available to those who are willing to face death.

Whereas Eros is about connecting in life, Thanatos is about death and rebirth. Physical death is an ending, through which each of us alive today must eventually travel, although we may hold different beliefs about what comes after physical human death. However, *psychological* death, which is letting go of what we identify with, is both an ending and a new beginning. As Krishnamurti (1996) points out:

> Death is the ending, the ending of my possessions, my wife, my children, my house, my bank account. ... [but in psychological death] where there is ending, there is a beginning ... there is an ending to attachment ... there is a totally different state of mind ... with the ending of that burden there is freedom. (pp. 331–332)

In a culture that denies the reality of death and the temporary nature of everything, many of us unconsciously identify with the things we think we can control, such as beliefs, wealth, power, and fame. Unfortunately, such identification easily leads to attitudes and behaviors that ultimately harm myself and the world, such as verbal attacks against others, ruthless competition, and even physical violence and war. The constant stream of "climate bad news" to which we all have access today can be a reminder to us that we are living lives of denial and bring us symptoms of depression and anxiety that we may want to label *crisis fatigue*. Actually, perhaps we are not actually as tired of crises as much as we are tired of hiding from the realities of life and death.

The pain we all feel when we read or hear about bad climate news can actually help us become aware of what we humans have done and are doing to ourselves and to the Earth. The pain we feel when we read the news is "real" pain—the core suffering, not the suffering we feel when we try to escape from reality. As we have discussed, it is true that humanity has been gradually harming our natural environment, and that harm has been accelerating over the past centuries. And our destruction of the environment will lead to the death of increasing numbers of human beings and other living things.

In the Sciences Resources 4.1 box, we see examples of the scary headlines that are presented to us in our daily news. When we read such news, we can pay attention to how we react inside. When we deny death, suppress loss, and try to minimize Thanatos, we may tend to avoid bad news and, in our avoidance, become more self-destructive and also harmful toward others and our world. In his text *Age of Anger: A History of the Present* (2017), Mishra writes about how chronic anger has become more prevalent in cultures across the globe. He uses the term *resentiment* to describe the ongoing resentment many people have today toward those whom they perceive as having a disproportionate share of dwindling global resources and misusing their power.

SCIENCE RESOURCES 4.1 Examples of difficult climate news

Scientists recently recorded the hottest summer on record across the Northern Hemisphere and linked these record temperatures to fossil fuel emissions. Average July and August temperatures were estimated to be about 1.5 degrees warmer than preindustrial levels (Paddison, 2023).

IMG 4.4

Recent studies now show that there has been significant and worldwide damage done to ecosystems across the world by wildfires. Over 1,000 animal and plant species are threatened by extinction from these fires (*The New York Times*, 2023).

Climate change can have "lifelong impacts" on the mental health of our youth. Children are susceptible to climate change impacts. These impacts, which can be irreversible, include damage to fetal development, cognitive and emotional skills, and physiological development. Children who belong to minoritized populations may be particularly at risk (American Psychological Association, 2023).

Although it is an absolute fact that there is rising inequality of wealth, power, and influence across much of our world—and, therefore, we understand why many people feel resentiment—as helping professionals, we also understand that longstanding anger by itself may not change anything in the world except cause more pain for those who are already suffering. Indeed, if our human anger was so good for the world, then it would have already helped, since anger has never been in short supply. Instead, we can teach our community members how to use their resentiment constructively, to enhance human cooperation, equality, and equity. It may not always be easy, but it is possible for us to heal and transform our expression of Thanatos so that we can move from hate and destructiveness toward acceptance of loss and death and a new desire to support the well-being of all people. We do this because we realize that loving kindness is much more effective than violence in creating the transformations we all probably actually want.

Becoming Friends With Eros: Bringing Eros Into the Light

When I become friends with Eros, I become conscious of the fact that I am a living being with a body who is connected with the universe. I become conscious of the extent to which I can experience and celebrate these wonderful connections I have with other people, other living things, and the universe in which we all live. I realize the importance of love in life and how love for self and others can give my life meaning and purpose.

When people ask how they can learn to love themselves more, I often suggest that they start by becoming conscious of how they actually feel about themselves now. The term *self-hatred* can be used to include the interrelated feelings of shame, inadequacy, self-loathing, and low self-esteem that most of us have, at least to some extent, from time to time. When I see the extent to which I may currently have self-hatred and the unnecessary suffering that it causes myself and others, this awareness can help me begin to make a transformation toward a new relationship with Eros, toward a greater self-love.

As John Bradshaw (1988) taught us, our shame heals as we learn to accept ourselves, especially our own perceived limitations. As we love ourselves more, we may find that we do not need as much to seek out the substitutes for love and joy that we have learned to habitually chase after, including substances such as alcohol and other drugs, distractions such as social media and screen time, and other common addictions, such as excessive work or constant busyness. Instead, I might find that I can take the time to notice and appreciate my life more in the here and now. And as I become friends with Eros, I may find that I am no longer as jealous of other people because I feel that I have enough and, more importantly, am enough.

Becoming Friends With Thanatos: Bringing Thanatos Into the Light

Thanatos is a teacher. When I become friends with Thanatos, I can become more conscious of the mysteries of life and death and of deep space and deep time. How was the universe created? Why are humans on the planet? Why am I here? How should I live my life? These are all questions that human beings may wonder about when we contemplate and accept the realities of life and death.

When I become friends with Thanatos, I may also become conscious of the extent to which I already engage in self-harm, suicidal, and other destructive behavior. I realize that there are probably "parts" of me that I actually need to let go of, that need to die, usually because they cause harm to myself and others. As I let them go, I may experience a "rebirth" in which my capacity to love and be loved starts to improve. This consciousness may include a reverent awareness of how I myself actually participate in co-creating the climate crisis. I may realize that I no longer want to participate in the destruction of my world because I see how it causes harm to myself, other people, other living things, and the ecosystems that keep us alive.

Thanatos can also teach me to become more conscious of any resentment and hostility I may feel toward myself, other people, other living things, and the ecosystems that support all life. And as I understand my own self-hatred, I can also learn how it is related not only to self-destructive behavior but also to my willingness to help destroy other beings and the ecosystems on the planet we all share.

In addition, since I know that when I die, there will be an ending of what I identify with in my life on Earth, I am willing to disidentify while I am still alive and let go of my attachments to what is not me. What does it mean to *disidentify*? *Identification* comes from the French word *identité*, which means to "regard as the same." Therefore, when I identify with a belief or object, I consider it the same as me. As we have discussed, identification can lead to violence because when I perceive threats to my beliefs or possessions, I can feel justified to attack those who seem to be attacking me. As I disidentify, however, I can become conscious of my identifications and thus can become free of my attachments.

In fact, an essential part of our well-being may be associated with our willingness to die, in a sense, every day and every moment. Building on Eastern wisdom traditions, in his *Here I Am, Wasn't I? The Inevitable Disruption of Easy Times*, Sheldon Kopp (1986) writes about how there is a constant death and rebirth in ordinary time. Each of us thus have the daily opportunity to let go of the present moment and move into the unknown experience of the next moment. In a sense, therefore, we can also disidentify from the stability and predictability that we all seem to crave, which does not actually exist in our lives. As Pema Chödrön (2022) has taught us, we can learn how to do deal with the "big death" at the end of our life by practicing letting go in every moment. From Pema's perspective, our lifelong spiritual work can be thought of as a preparation for a future graceful and fulfilling death.

During the years I worked at the Indian Walk In Center in Salt Lake City, I was told by Ute and Navajo people that the phrase *today is a good day to die* was used in a variety of contexts and variations by some tribes in what is now North America. One story is that Crazy Horse used this expression in his talk with other warriors before the Battle of Greasy Grass (known as the Battle of Little Big Horn in the White world). I was told that this phrase means that I have lived my life fully and therefore am prepared to die today if that's what is meant to be. This idea seems congruent with the meanings of Eros and Thanatos that we have been exploring. When I become friends with Eros, I want to live my life fully, and acceptance of death is associated with the idea of becoming friends with Thanatos.

Eros, Thanatos, and Our Climate Crisis: A Story of Healing and Transformation

Creating New Stories

It has been said that the universe is made up of stories, not subatomic particles. One way to use archetypal psychology to work with our consciousness and our climate crisis is to use archetypes to create new stories about our history, bridging from the past through to the future. Such stories can serve to inform, challenge, and inspire us to reach our potential as a species. The following example builds on the Eros and Thanatos archetypes as well on the archetypal concept of ego to write a story. Please simply consider this as just one example of how to do an exercise that we could all consider doing, as we co-create new ways of feeling and new rituals of participation in our lives. Also, please note that I have written another story in Chapter 15 that is designed as a creation myth that tells a story of how consciousness is related to human, origins, survival, and evolution.

A New Story of Our History

Exactly how Creator brought the first life into the world is still a Great Mystery. Perhaps life arrived here on Earth multiple times. We know all life on our planet evolved from a simple organism that lived somewhere in the universe about 3.5 billion years ago, maybe much earlier than that. Having the same great-grandmother, this first organism, every living thing is connected with all other living things as well as with the universe that sustains us.

All life is able to reproduce, and all living things also are designed to die so that life can make room for new generations to grow and progress. Recognizing both life and death, we say that Creator brought into the world Eros, the drive toward life, and Thanatos, the drive toward death.

Even these simplest living things on what we now call Earth were given the complementary gifts of Eros and Thanatos. Blessed with Eros, all life enjoyed being alive and were able to reproduce and evolve. Blessed with Thanatos, the first living things lived in a universe that was constantly changing, from moment to moment; they lived fully *because* they also accepted loss and death when it inevitably came. All living things cooperated together in ecological systems to draw nourishment from the water, land, and atmosphere and from each other's bodies. For every species, death was the way we made room for the many descendants who were destined to follow.

Eros and Thanatos were expressed through evolution; many species arrived, enjoyed life, and then disappeared, making way for new species to arrive and do the same. Many living things surrendered their bodies to other living things who needed to consume them to survive. Our early Earth ancestors, the one-celled living things, prospered by themselves for millions of years and helped create an environment on the planet for the other forms of life to come.

Some of the early multicell creatures evolved into the invertebrates. The first mammals, which arrived much later, struggled for success until most of the dinosaurs went extinct about 65 million years ago. Just as individual life forms were born, lived, and died, so did species come and go. Over many millions of years, the average invertebrates lasted from 5 million to 10 million years before they went extinct, the average mammals 1 million to 2 million years. The time had come for the first "modern" humans to emerge on the planet. Gradually, our species, *Homo sapiens*, evolved from the later mammals, appearing about 315,000 years ago in what we now call Africa.

Now Ego was the child of Prometheus, the god of anticipation and crafty counsel who helped oversee the creation of the first humans, and Pheme, the goddess of fame and gossip who had the power to give people celebrity or social rejection. Ego was jealous of the power and status of Eros and Thanatos, did not want humans to worship them, and retaliated against them by giving the first humans the innate need to identify with their beliefs, wealth, power, and fame.

This desire for identity and identification was a mixed blessing. Although there were many challenges, these first humans were successful in surviving and gradually migrating across the world. The humans were able to use their brains and egoic self-interest in new ways that enabled them to outcompete all other living things. Our ability to put ourselves first helped us survive, and we learned how to control nature and compete with each other in our accumulation of wealth, power, and fame. Our numbers grew from about 4 million about 12,000 years ago, to about 1 billion people in 1900, to more than 8 billion people on the planet today. We "modern" humans have survived so far for only about one-third of a million years, which means that as typical mammals, we might expect to survive roughly another 660,000 to 1,660,000 years.

Ego caused mankind to think of ourselves as separate from and superior to all other living things. In some significant ways, of course, we are different. No other Earth species was able to communicate, create cultural transmission of knowledge, and invent new technologies like we have. We have become successful at populating the planet, dominating most other living things, and drawing resources from the ecosystems that sustain us. However, the other living things and ecosystems have paid a steep price for our "success," and our activities have resulted in the accelerated death of many of our sister and brother species. We have polluted our water, land, and atmosphere systems across our planet. The background extinction rate, the rate at which species die, was estimated to be at about one species extinction per 1 million species per year until humans took over the planet. Now, today, because of human activities, that rate has exploded to a rate of about 100 to 10,000 extinctions per 1 million species per year.

Ego had apparently won the power struggle with Eros and Thanatos and now dominated the affairs of humankind. However, there was now much unnecessary suffering. Despite our increasing dominance over other living things and the ecosystems that sustain life, humans failed to confront the biggest threat to our own survival and well-being, which was the power of Ego in our own consciousness.

Unfortunately, in our desire to accumulate wealth power and fame, we began to lose our awareness of the interconnection we have with the other living things and ecosystems that birthed us and sustain us. We lost our awareness of Eros, the god of life, and chased instead the shadow state of unconscious Eros, in which we sought after such substitutes for Eros as money, status, and increasing control over other people, other living things, and what was left of the natural world. We also lost our awareness of Thanatos, the god of death, as we resisted the inevitability of death and the need to make way for our descendants, as our ancestors made room for us. As we pushed Thanatos into shadow, we substituted self-harm, suicide, and destructive behavior for our former acceptance of life and death. We lost the awareness that other living things seem to have, that enlightened self-interest is about the interconnectedness and common well-being of everything.

With our identification with belief, wealth, power, and fame, we began endless cycles of conquest and war that has continued to today. Instead of letting go and disidentifying from those things that are not who we truly are, we clung to our beliefs, went to war with those who disagreed, and continued to destroy the ecosystems that keep us alive. We became increasingly willing to take every bit of the natural world we could take to feed our increasingly addictive need for our own narrow understanding of self-interest.

More and more people all over the world became depressed and anxious in reaction to the deterioration of our beautiful planet, as we all saw accelerating species extinction, extreme weather, destruction of habitat, and pollution of our land, water, and atmosphere. People who belonged to minoritized populations often suffered the most, as they were unable to afford to live in areas where the climate was less dangerous and were often unable to purchase, for example, the air conditioning, quality food, or drinkable water that they needed to stay healthy. The wealthy classes were often able to ignore the climate crisis longer than the poor, and conflicts, resentments, and wars between wealthy and poor populations increased.

There were many who at first reacted by going into denial or minimization of the climate crisis, and their fears were encouraged by some politicians who wanted power and by some corporate leaders and investors who still sought to profit from the extraction of coal, oil, and natural gas resources and from industries that emitted the greenhouse gases and other pollutants that were slowly killing us.

Ironically, Ego did not become happier as it gained more power and, in fact, went into deep despair and suicidal thinking. Eros and Thanatos could see this and had compassion for Ego. They offered Ego a role in helping them co-direct human evolution. Together, they would help humanity become conscious of Ego and of the benefits and risks involved in letting self-interest dominate human evolution. Instead, they crafted a plan for the enlightened self-interest of humankind. Ego agreed to make peace and cooperate.

So, it came to be that, because of our physical, emotional, and spiritual suffering, more of humanity became willing to face the challenges that we now face. Many people became painfully aware of how their children, grandchildren, and other deep-future descendants would have to live in steadily deteriorating conditions. Increasing numbers of people became climate refugees, forced to leave their homes because of heat waves, fires, flooding, or drought. As governments saw that the majority of us wanted to make the changes that were necessary to reverse global warming, those changes were finally made.

An increasing feeling of cooperation was fostered worldwide, as people realized that we had to work together to survive the climate crisis as well as such related crises as preparations for war, pandemic, and mass-casualty terrorism. Wise leaders made sure that no populations had to suffer disproportionately as changes were made. For example, coal industry workers as well as mechanics who repaired vehicles and other machines fueled by fossil fuels were given assistance in finding new employment and even new locations to live in if they wanted such opportunities. At a more macro scale, poor countries in the world were given assistance in creating renewable energy for their populations because wealthy nations increasingly realized that offering such assistance ultimately served the highest good for all humans, including those living in wealthy nations.

As people realized that future generations would benefit from reduced population, an effort was made to do just that. This process was difficult, because all populations naturally wanted to feel that the process was fair, and there was initially a deep lack of trust in many populations, and some economists howled, but eventually the world population was reduced to 1 billion people, and the quality of life improved dramatically for everyone on the planet. Humanity finally saw that the well-being of every individual human being is interconnected with the well-being of every living being. All minoritized populations were welcomed back into the pantheon of life on Earth, and the well-being of our ecosystems were once again seen as the responsibility of everyone and the birthright of every living things.

Human beings rediscovered a reverence for all life on Earth. New and compassionate ways were found to care for all life and provide clean food, soil, water, and air for every living thing. Children were raised to live with a faith-based activism in which they refused to go into despair, self-hatred, and destructiveness. Instead, they chose to live lives dedicated to understanding and supporting the highest good, which is to say that they learned the wisdom early to put the collective well-being of all life first in the value hierarchy of their lives.

As space travel and settlement became more accessible to everyone, increasing numbers of people immigrated to other planets, solar systems, and galaxies. Our newfound ability to cooperate to solve the climate crisis enabled us to also collaborate to travel to and inhabit outer space. However, the lessons of conservation and sustainability were never forgotten and were applied to the new extraterrestrial locations human beings started to live in. The human species had survived the climate crisis era, and all living beings on the Earth began to truly live in more peace and prosperity. .

QUESTIONS FOR REFLECTION

1. Were you familiar already with the idea of archetypal psychology before you read this chapter? If so, what are your thoughts and feelings about it? If not, what are your first reactions after reading the chapter?
2. Reflect on a time when you have thought about suicide and perhaps even felt suicidal. What kept you alive? Was there a part of you that actually needed to die?
3. Can you relate to the archetypes of Eros and Thanatos? Can you see them in yourself and in other people? Please explain.

4. Do you agree that Eros and Thanatos offer us ways to think about our climate crisis? Please explain.
5. After reading "A New Story of Our History," what was your first reaction? How did you like the ending? Does it seem realistic or even possible? Please explain.
6. If you would prefer to write another ending to the story or even a completely new story, what would it be? If you liked the ending, please explain why.

REFERENCES

American Psychological Association. (2023). *Impacts of climate change threaten children's mental health starting before birth*. October 11, 2023. https://www.apa.org/news/press/releases/2023/10/climate-change-children-mental-health

Bradshaw, J. (1988). *Healing the shame that binds you*. Health Communications.

Brown, N. O. (1959). *Life against death: The psychodynamic meaning of history*. Random House.

Brown, N. O. (1966). *Love's body*. Random House.

Chödrön, P. (2022). *How we live is how we die*. Shambhala Publications.

Frankopan, P. (2023). *The Earth transformed: An untold story*. Knopf.

Freud, S. (1930). Civilization and its discontents. In E. Jones (Ed.), *The international psycho-analytical library* (7th impression). Hogarth Press.

Kli, M. (2018). Eros and Thanatos: A nondualistic interpretation: The dynamic of drives in personal and civilizational development from Freud to Marcuse. *The Psychoanalytic Review, 105*(1), 67–89. http://dx.doi.org/10.1521/prev.2018.105.1.67

Kopp, S. B. (1986). *Here I am, wasn't I? The inevitable disruption of easy times*. Bantam.

Krishnamurti, J. (1996). *Total freedom: The essential Krishnamurti*. HarperOne.

McGilchrist, I. (2009). *The master and his emissary: The divided brain and the making of the Western world*. Yale University Press.

Mishra, P. (2017). *Age of anger: A history of the present*. Farrar, Straus and Giroux.

The New York Times. (2023). Transformed by fire. October 17, 2023, D1, D8.

Paddison, L. (2023). *The world has just experienced the hottest summer on record—by a significant margin*. CNN, September 26, 2023. https://www.cnn.com/2023/09/06/world/hottest-summer-record-climate-intl/index.html

Shakespeare, W. (thought to be written between 1599 and 1601). *Hamlet*.

Taylor, L. (2021). The beauty and brilliance of shadow work. *Psychology Today*, September 28, 2021. https://www.psychologytoday.com/us/blog/reflections-neurodiverse-therapist/202109/the-beauty-and-brilliance-shadow-work

Vince, G. (2020). *Transcendence: How humans evolved through fire, language, beauty, and time*. Norton.

Yalom, I. (2008). *Staring at the sun: Overcoming the terror of death*. Jossey-Bass.

Credits

IMG 5.1

Despair, Hope, and Faith-Based Activism

My friends, do not lose heart. We were made for these times. I have heard from so many recently who are deeply and properly bewildered. They are concerned about the state of affairs in our world now. Ours is a time of almost daily astonishment and often righteous rage over the latest degradations of what matters most to civilized, visionary people.

—CLARISSA PINKOLA ESTES (2023)

Keep the faith.

—JIMMY CARTER (1995)

My mentors made a big difference in my life. Perhaps like many people in my generation—and perhaps in most modern generations—I entered adulthood with an unsatisfied mentor hunger. I found inspiration from a number of people I met during my undergraduate college years, mostly professors and activists, who increased my interest in working in a university setting. They saw the potential in me; they saw what Robert Johnson (2018) called the "inner gold" inside me, before I was able to see it in myself. I wanted to do the same someday for others.

I also was fortunate to read the books written by other mentors who also made a positive difference. When Kenneth Keniston published *Young Radicals: Notes on Committed Youth* in 1968, I was looking for new ways to think about my life meaning and purpose. His book inspired me to look toward service to others as a life path. He wrote about how young people who were devoted to a cause could both benefit themselves and their world; therefore, a commitment to a cause could be healthy for both my outer and inner worlds.

People in my generation grew up in fear of a nuclear war. As grade-school children in a Chicago classroom, we were encouraged to get under a desk and cover our face during nuclear attack drills. As we crouched obediently, we also realized how futile it was to practice these drills, since we knew we would all probably be vaporized in a real nuclear attack. When Joanna Macy published *Despair and Personal Power in the Nuclear Age* in 1983, I was grateful that someone was there, thinking about how we all felt in that era of nuclear vulnerability, and the idea of hope was very appealing. Macy's "active hope" was about becoming active participants in co-creating the world we hope for. Former President Barack Obama (2014) also often discussed the importance of hope. I liked how Obama seemed to think of hope as the opposite of fear, fatalism, and cynicism. Even when his political opponents made fun of him, he continued to talk about the power of hope.

Marianne Williamson has been someone who has always seemed willing to be a light in our time of spiritual and political darkness. I like how she admonishes people to "get real and authentic and look in the mirror." Although they always put her on the far end of the line of podiums during the national debates, she still persevered, discussed what was actually going on, and refused to express or promote hate. Although people often viewed her like a child who maintained silly fantasies of love and peace, she seemed to be one of the few people acting like an adult, who was willing to talk honestly about what was really happening in the world. I agree with her that the majority of people in the world actually do want the same things, including love, deep peace, and inner peace, and that it is the responsibility of each of us to do what we can to give that majority a voice in decision-making.

I was also influenced by insights from Eastern philosophy that challenged me look more closely at hope. I was most influenced by the writings of J. Krishnamurti (1996), who reminded us that hope could function as a defense mechanism in which we grasp for an imagined future in our desire to escape the suffering we have in the present moment. He taught that when we look outside of ourselves, to any external authority, for an answer, we will ultimately become disappointed; however, if we engage in true meditation, we are capable of seeing things as they actually are.

So, is hope the "medicine" for the despair that so many of have felt throughout our lives, or does the way we teach hope often result in doing more harm than good?

The work of creating this book project chapter challenged me to reflect again on despair and hope, and having mentors who disagreed challenged me to think for myself. Many of my colleagues in academic social work currently emphasize the installation of hope as a fundamental goal of psychotherapy. Yet, I have continued to feel that Krishnamurti was right, that the attitude of hope, as commonly taught in our Western culture, was not necessary for the transformation of mankind and was too often harmful to ourselves, other people, and the wonderful world we want to protect.

A Loss of a Sacred Landscape

I try to get out every summer to spend times out in the desert alone, to do some wandering and self-reflection. About three years ago, I drove down to one of my favorite places to spend some time there. As I drove down the old mining road near where I wanted to camp, with dust flying around my truck, I noticed some new trails cut into the earth off the side of the road. Pulling up near the ancient pinyon pine that I consider one of my old friends, I saw that the whole campsite had been trashed by off-road vehicles. There were newly cut wheel tracks crisscrossing the whole area and destroying the cryptobiotic soil crust, and there were fresh wounds on the trunk where the still-living branches of "my" tree once grew, the branches probably used for firewood in the ring of rocks, now surrounded by beer cans and other waste scattered on my holy ground.

As a friend of mine told me after having a similar experience in a forested area he loved near his home in northern Utah, it was as if someone had driven through my neighborhood church, plowing through the walls, pews, and other sacred objects and structures. One of my most sacred landscapes had been vandalized.

I realized that my grief about the damage to the area was complicated by feelings of fear and anger that some of our fellow human beings will continue destroying all of our last surviving relatively untouched wild areas. I have wondered about the people who caused the destruction. Why would they raze an area that was probably as beautiful to them as it is to me? Could the unconscious shadows of Eros and Thanatos be operating in their lives, as they sometimes do in mine? And what would a dialogue with some of those folx be like? Could we talk together about how we all feel and think about the climate crisis in a civil way and find some common ground? Could we heal together the despair both they and I probably feel about what is happening to our remaining natural resources?

In Chapter 2, we looked at how important it is to allow ourselves to feel our feelings about the climate crisis and about our lives in general. We started, for example, a discussion about sadness, which is the natural reaction to loss, and how healthy it can be to grieve our losses and anticipatory losses. What follows in this chapter is my attempt to develop a theory of faith-based activism that can be a useful foundation for our work with climate crisis and perhaps with any crisis.

The Benefits and Shortcomings of Hope and Despair

What is hope, and what is despair? How can these attitudes benefit us, and how might they cause us harm?

The etymology of these commonly used words are shown in Table 5.1. As you can see, the word *hope* has origins in a religious-based confidence in God's plans for our future. An examination of the literature shows that hope has been used in many different ways, with various definitions and meanings, including:

1. A specific goal-oriented wish
2. An acceptance of today's realities combined with determination
3. A feeling that motivates me and helps me make sense of the situation
4. An attitude that can help empower minoritized and marginalized populations
5. A motivation associated with action
6. A duty expected of people
7. A virtue of some people
8. An attitude that brings psychological and social benefits (Green, 2018; Brei, 2016)

In contrast, the word *despair* is usually used to describe a loss or lack of hope. There seems to be a general agreement in the literature that chronic despair is not helpful to human beings (Brei, 2016).

TABLE 5.1 Etymology of Key Words in Chapter 5

hope (n.)	From the late Old English *hopa*; "confidence in the future," especially "God or Christ as a basis for *hope*," from *hope* (v.). From circa 1200 as "expectation of something desired"; also "trust, confidence; wishful desire"; late 14th century as "thing hoped for"; also "grounds or basis for hope." Personified since circa 1300. Related: *Hopes*.
despair (n.)	From circa 1300, *despeir*, "hopelessness, total loss of hope"; from Anglo-French *despeir*; Old French /*despoir*, from *desperer* (see *despair* [v.]). The native word was *wanhope*.
faith (n.)	From the mid-13th century, *faith, feith, fei, fai*, "faithfulness to a trust or promise; loyalty to a person; honesty, truthfulness"; from Anglo-French and Old French *feid, foi*, "faith, belief, trust, confidence; pledge" (11th century); from Latin *fides* "trust, faith, confidence, reliance, credence, belief"; from the root of *fidere*, "to trust"; from the Proto-Indo-European root **bheidh-* "to trust, confide, persuade." For sense evolution, compare *belief.* It has been accommodated to other English abstract nouns in *-th* (*truth, health*, etc.).
activism (n.)	From 1920 in the political sense of "advocating energetic action"; see *active* + *-ism*. Earlier (1907), it was used in reference to a philosophical theory. Compare *activist*.
attitude (n.)	From the 1660s, "posture or position of a figure in a statue or painting," via French *attitude* (17c.); from Italian *attitudine*, "disposition, posture"; also "aptness, promptitude," from Late Latin *aptitudinem* (nominative *aptitudo*; see *aptitude*, which is its doublet). Originally from the 17th century; a technical term in art; later generalized to "a posture of the body supposed to imply some mental state" (1725). The sense of "a settled behavior reflecting feeling or opinion" was in use by 1837. The meaning "habitual mode of regarding" is short for *attitude* of mind (1757). Connotations of "antagonistic and uncooperative" were developed by 1962 in slang.

Source: All quotations in Table 5.1 were adapted from Online Etymology Dictionary (https://www.etymology.com/).

Today, many theorists view hope as a fundamental element necessary in the professional helping process. Respected therapist and author Irving Yalom (Yalom & Leszcz, 2020) included the "installation of hope" in his list of the 11 key therapeutic factors of psychotherapy. Fraser (2019) states that "all therapies that work offer a *frame, rationale, or explanation* of clients' problems, that *implies a direction* for resolution, and *engages clients' energy to go that way.* In a nutshell, they build a framework of hope, and hope is claimed to be correlated with success in therapy."

If hope can be helpful, then what kind of hope, and in what circumstances? And in what circumstances can hope become harmful?

Starting long before we invented our Western psychotherapy, Buddhist teachers have examined these issues for centuries, looking at hope in a deep and nuanced way and examining the potential for both harm and benefit. In a wonderful article in *Lion's Roar,* three Buddhist teachers, Ayya Yeshe, Sister Clear Grace, and Oren Jay Sofer (2023), have a conversation about hope:

Sofer begins by commenting:

> What we might call "ordinary hope" directs our longing for happiness in an unskillful way. It places our well-being on an uncertain, imagined future beyond our control, thereby feeding craving and fixation. When the wished-for outcome isn't realized, we are crushed. ... Dhamma practice channels our longing for happiness, harmony, and equity in a skillful way. This begins with *saddha*, most frequently translated as "faith" or "conviction." *Saddha* [emphasis added] refers to one's aspiration and confidence in the path. It is the intuitive sense that there is something worthwhile about being alive, that inner freedom is available for each of us.

Sister Clear Grace responds:

> ... The Buddha teaches us that there are three kinds of people in the world: The hopeful, the hopeless, and the one who has done away with hope. ... Hope acquired through direct experience gives us insight into change, rather than just the wanting of change. This wise hope can allow us to see things as they are—that nothing is inherently permanent or fixed. The Buddha directs us to a path that is wishless or without expectation. It is from this very space that we are then able to create and be the very hope that we wish to see.

Finally, Ayya Yeshe concludes:

> ... what if we look at hope as something different from desire? What if we acknowledge that we are not enlightened yet, and that hope as resilience—a long-term commitment to practice and social justice and compassion, equanimity, and watering the seeds of joy and happiness in ourselves—is a necessary part of the courage, strength, and endurance needed to become bodhisattvas, to become enlightened, and to create a more just world?

As Krishnamurti (2022) has observed, hope should be seen as a defense mechanism, an escape from the suffering we have in the present moment, and a reaction to despair. He reminds us that our hope in other people, such as teachers, therapists, or politicians to save us, for example, will always result in disappointment and further suffering. Krishnamurti suggests that we face every crisis without fear and without hope:

> We are born with hope and we take it with us to death. ... Hope is the desire for the continuation of that which has been pleasant, of that which can be improved, made better; and its opposite is hopelessness, despair. We swing between hope and despair. ... To live happily is to live without hope. The man of hope is not a happy man, he knows despair. ... Is there not a state which is neither hope nor hopelessness, a state which is bliss? After all, when you considered yourself happy, you had no hope, had you? ... First see the truth of hope and hopelessness. Just see how you have been held by the false, by the illusion of hope, and then by despair. ... Your problem is unhappiness, and to understand it there must be freedom from all other problems. Unhappiness is the only problem you have; don't become confused by introducing the further problem of how to get out of it. The mind is seeking a hope, an answer to the problem, a way out. See the falseness of this escape, and then you will be directly confronted with the problem. It is this direct relationship with the problem that brings a crisis, which we are all the time avoiding; but it is only in the fullness and intensity of the crisis that the problem comes to an end. (2023, pp. 1–2)

Keep the Faith

I have always admired former President Jimmy Carter for his integrity, public commitment to social justice, and personal lifestyle. In the 49th United States presidential election in 1980, Ronald Reagan and George H. W. Bush defeated incumbents President Carter and Vice President Walter Mondale. I was bummed and wrote Carter a letter stating how much I admired him and how disappointed I was that he lost the election. He took the time to reply with a brief handwritten note that stated simply, "Keep the faith."

President Carter is a very religious man and has a very strong personal and active faith. *CBS Mornings* (2023) offered a recent interview of Jimmy's pastor, Tony Lowden, who praised the former president at the time when he entered hospice care. Pastor Lowden talked about his friendship with Carter and about Carter's faith and lifelong commitment to civil rights and other social justice issues.

People from many backgrounds, religious or not, can live a faith-based life. Michael Jackson, for example, whose religious beliefs apparently changed and developed during his short life and remain unclear (Celebrity Beliefs, 2017), performed his song "Keep the Faith" on his 1991 LP *Dangerous*. The song is a moving piece about the importance of finding the power inside of yourself to follow your dreams. We can think about faith, therefore, as an attitude toward oneself and the world that may or may not be associated with a person's familial, cultural, or religious backgrounds.

The word *faith* implies having trust and confidence in what we hope for. In this chapter and in this text about climate change, I would like to propose we use the phrase *faith-based activism* to describe a hope for a better *future* that both inspires and includes *actions* in service to the world in the *present*. The etymology of the words *faith* and *activism* are included in Table 5.1.

Models of Hope Practice That Include Action

Following are three models for a practice of hope. Each addresses the complex mix of potential benefits and dangers of hope that we have started to examine in this chapter.

Radical Hope

In *Radical Hope: Ethics in the Face of Cultural Devastation*, Lear (2006) writes about the story of the last days of freedom for the Crow tribe and about their "last great chief" Plenty Coups. Lear recounts how the Crow had for centuries enjoyed and guarded their way of life. The tribe had a love of honor and especially valued warriors who counted coup in battle. *Counting coup* was a way of winning prestige against an enemy by contacting an enemy with a hand, weapon, or coup stick without being harmed or causing harm, especially when such an act caused the enemy to be demoralized or surrender. Plenty Coups had earned the greatest prestige of all the warriors.

As White people began arriving, their killing of buffalo began to shrink the herds, and many Indian people died from the new conflicts and the new illnesses carried from Europe. Then, the surviving Crow were sent to a reservation. The tribe was experiencing what could be called a cultural catastrophe, and after the tribe's move to the reservation, Plenty Coups observed that the hearts of his people seemed to fall to the ground and could not be lifted up again.

However, Plenty Coups had a radical hope that someday things would be better again. He did not give up. While living with his people on the reservation, Plenty Coups tried to find a middle ground in his conflicts with Whites. He tried to find a genuine, positive, and honorable way of going forward and surviving the catastrophe.

Believing that his people would get their culture back someday and staying humble in the face of uncertainty, Lear states that Plenty Coups had a "radical hope" that "was directed towards a future goodness that transcends the current ability to understand what it is" (p. 103). Coups's hope was connected with his quest for how to best live in the present and provided a basis for his holding courage for his people in an uncertain and challenging time. The destruction of the culture was, he recognized, offering a route to rebirth. Coups ultimately successfully led his people in finding ways to be successful in the White people's world.

Active Hope

Joanna Macy (Macy & Johnstone, 2012), whom we introduced earlier in this chapter, has written and taught about hope and despair throughout her career. Having recognized the "mess" that the world is in, including threats of war and the climate crisis, her concept of active hope is an empowerment approach that draws upon multiple modern and ancient wisdom traditions. With co-author Johnstone, Macy advocates for a "Great Turning" in which humanity builds on our hope and participates actively in co-create a culture that sustains life. As Macy (2023a, p. 1) remarks, "Of all the dangers we face, from climate chaos to nuclear war, none is so great as the deadening of our response. ... Active Hope is waking up to the beauty of life on whose behalf we can act. We belong to this world."

Macy (2023b) also writes:

> At the heart of this book is the idea that Active Hope is something we do rather than have. It involves being clear what we hope for and then playing our role in the process of moving that way. The journey of finding, and offering, our unique contribution to the Great Turning helps us to discover new strengths, open to a wider network of allies and experience a deepening of our aliveness. When our responses are guided by the intention to act for the healing of our world, the mess we're in not only becomes easier to face, our lives also become more meaningful and satisfying.

Hope for the Hopeless

A third example of a practice of hope comes from 13th-century Japan. Higa (2023) writes about the life work of Shinran Shonin, a Buddhist monk living during the Kamakura Period in Japan. Shinran was the founder of what ultimately became the Jōdo Shinshū sect of Japanese Buddhism, which became the largest school in Japan and eventually also one of the oldest in the United States. Higa states that Shinran was very honest about himself and the human condition and quotes him as saying, "Hell is decidedly my abode whatever I do." Shinran felt that it is hard to be a human and that we are quite limited in our ability to reach spiritual maturity, or enlightenment. The Japanese used the word *Icchantika* to describe this human condition, which translates as the "hopeless one." However, Shinran taught that there is "hope for the hopeless" if we work to understand and accept our own "spiritual imperfection." He taught that we need to learn from the mind of Buddha, become aware of our true nature, and accept our limitations as well as our strengths.

What Is Faith-Based Activism?

As introduced earlier, we will use an approach to hope called *faith-based activism* in this text. In this section, we will examine what the components of such an approach look like.

Faith-Based Activism Combines Faith With Action

We start with the idea that faith-based activism is part of the *climate change* we need in both human consciousness and behavior, which includes the cultivation of faith, reverent awareness of oneself and the world, and a commitment to engage in skillful activity. All these components are useful and necessary for us to work together on the climate crisis.

Faith is a positive attitude toward the world and, ultimately, a feeling of trust that the universe is a friendly place. As we can see in Table 5.1, *attitude* comes from a root form that meant "a mode of regarding." Albert Einstein (Goodreads, 2023) was quoted to have said, "The most important decision we make is whether we believe we live in a friendly or hostile universe."

It is fascinating that he thought that our attitude toward the universe is so essential and central to our lives and that he saw this kind of faith as a decision.

In fact, one of the few things over which human beings may have some control is how we choose to regard ourselves and our world. There is evidence that this ability to choose our own attitude is a foundation of successful human relationships and, ultimately, human happiness (Glauser, 2022). Each of us is challenged in every moment to make reaffirm this climate change in our consciousness, as we strive to see ourselves and the world accurately and with friendly eyes.

My faith enables me to persevere, especially when the world seems cold and unfriendly. *Perseverance* is defined as "finishing what one starts; persevering in a course of action despite obstacles; 'getting it out the door'; taking pleasure in completing tasks" (Schaffner, 2020, p. 1). Researchers have found that there is a strong link between perseverance and resilience when people follow their personal passion and make a commitment to a practice of service to others (Schaffner, 2020). Such hope, both bold and determined in the face of uncertainty, was modeled by Dr. Martin Luther King, Jr., who said, "We must walk on in the days ahead with an audacious faith in the future" (The Spiritual Life, 2023).

We Were Made for These Times

The leading quote in this chapter, from Clarissa Pinkola Estes (2023), begins, "My friends, do not lose heart. We were made for these times." I love that quote. Faith-based activism builds on this foundational attitude—that those of us alive on the planet today have the ability and perseverance to deal with climate change effectively. We can rise to the challenges given to us.

In her presentations, Williamson (2023) often likes to echo those words from Estes. It is easy to "lose heart" in the times in which we live, with the many global crises seeming to converge on us right now. Williamson reminds us that whether we consider ourselves religious, spiritual, agnostic, atheist, or any other category, we can all operate out of a belief in our ability to meet the challenges we are given in our lifetime.

Activism and Consciousness

Activism, as shown in Table 5.1, comes from a root form that refers to "advocating energetic action." We will view *activism* in our text as a climate change in our relationships with our world. In such a change, we make a choice every day to express loving kindness toward self, other people, other living things, and the world in which we live.

All our activism is best built on the foundation of our consciousness. The process of faith-based activism is informed by our consciousness and our ability to see things the way they really are. When we see ourselves and the world accurately, we know what to do. In Sections 2 and 3 in the text, we will be exploring ways to actively model and support climate activism with other human beings, other living things, and the ecosystems that support all life on Earth.

Faith-Based Activism Is Based on Love and Empowerment

Faith-based activism is based on love, even (and perhaps especially) for those who disagree with us. As Dr. King (Reader's Digest, 2023) said, "Darkness cannot drive out darkness: Only light can do that. Hate cannot drive out hate: Only love can do that."

Dr. King often reminded activists that we cannot hope to positively influence those whom we still hate, but when we forgive those who have offended us, we may see their hearts and minds open up.

I find that forgiveness is often misunderstood by my students and clients. When I talk with them about forgiveness, they often angrily respond by saying something like, "Now, I am supposed to like those people and condone their misbehavior?" Forgiveness does not require me to like or approve of anyone's misbehavior.

Instead, I ask them to consider that all forgiveness means is that I choose to give up my anger, resentment, and hate because I realize it is in my own highest good to do so. If I choose to forgive you, I am not agreeing to necessarily like you, spend time with you, agree with your beliefs, or condone your behaviors. All I am doing is letting go of feelings that no longer serve me, because my anger, resentment, or hatred is poisoning me and gets in the way of my ability to help create positive change in the world I love.

Williamson (2023) also built on this idea of love in her career of writings and public presentations. She reminds us that love is not passive but rather always active. For example, we do not just discuss our love for other species and ecosystems; we also *do something* to help protect all life on Earth. And she also reminds that our moral outrage about the climate crisis does not by itself do much to protect Earth and Her creatures.

Williamson offers us additional wisdom. According to her, love sometimes needs to say no—such as with the mother who takes the sharp object away from the unknowing and vulnerable infant. Also, loving kindness does not mean that we fail to speak out when something is wrong. Faith-based activism is about love for self, family, community, country, and the Earth. Love is not blind; we see and strive to accept the imperfections in ourselves, our family, community, and country. I can speak to the imperfections and mistakes that I make and that others make and can still actively care for the well-being of all of us.

Self-empowerment is also associated with faith-based activism. *Empowerment* can be seen as the ability to understand, accept, and skillfully and nonviolently express our thoughts and feelings. In contrast, *power* usually involves an attempt to control other people by whom we may feel threatened, and such control usually includes psychological and sometimes physical threats and violence. An attempt to have power and control over others usually results in a reaction from others to seek power and control over us. *Self-empowerment* involves a shift away from any need I might have to control other people. As Adam Blanch (2018) writes, "the most powerful person in the world is the one who doesn't need to avoid their feelings. If I am prepared to feel my fear, feel my hurt and my shame and not turn away from the reality of my vulnerability, I become impossible to control and impossible to denigrate."

For Williamson, Dr. King, and many of us, we each have a *response-ability* to make politics an integral part of a meaningful life. In such work, we strive to support the empowerment of all voices in our world. Although the majority of us want positive change, many and probably most of us who have solutions to offer do not yet have political power, and many of those who currently hold political power in our world do not yet want to hear from those with solutions. We all, however, can strive to empower ourselves and help bring light into the darkness.

Faith-Based Activism Is Inclusive

An inclusive faith-based activism welcomes all participants, reaches out to all people, and considers the well-being of all living things on Earth. Perhaps Dr. King said it best: "If we are to have peace on earth, our loyalties must become ecumenical rather than sectional. Our loyalties must transcend our race, our tribe, our class, and our nation; and this means we must develop a world perspective."

We recognize that such a shift from identifying with our personal characteristics and beliefs towards perhaps identifying more with a global well-being will, of course, be challenging to many of us.

In our activism, we want to include as best we can the voices of all the people in the communities with whom we work. Although the climate crisis has global interventions, as we have seen, we will often work at a local level, and the concerns, traditions, and values of local people need to be recognized in the decision-making process. We know that addressing the climate crisis will require sacrifices from people, and these should be shared as equally as possible over time.

In addition, as we will explore further in later chapters, we want to strive to give other living things and our ecosystems a voice in policymaking. We can help represent the well-being of trees, wild and domesticated cattle, and desert streams, for example, in spaces in which relevant policymaking is happening.

Faith-Based Activism Faces and Builds on Both "Bad" and "Good" News

Many of my students, clients, and colleagues talk about how hard it is to have faith in a world of bad news. However, bad news can also actually help strengthen faith, and there is always also "good news" if we want to look for it. Earlier in the text, we provided a "Bad News" box; now, in the Science Resources 5.1 box, we illustrate some examples of good news during out climate crisis. If we look, we can always find positive climate news.

Psychologists have learned that our feelings of helplessness and despair can be healed and transformed through a shift in focus. Instead of focusing primarily on healing trauma, professional helpers can also focus more on working with clients to support them in identifying those things that they *can* do to help make life worth living (Seligmann, 2019).

Similarly, researchers have noted how the news we consume tends to be about human suffering and crisis and that such constant negativity can lead to increased desensitization, cynicism, and resistance to becoming actively involved in our world (Kelsey, 2020). A shift toward exploring news about people trying new solutions can help people regain faith and activism. One of my favorite journals, for example, is *Yes! Magazine* (https://www.yesmagazine.org/), which offers a buffet of new ideas and potential solutions in every issue. We could even postulate that adversity is the mother of faith and that we do not even need faith until we encounter challenges in life.

SCIENCE RESOURCES 5.1 Good news

IMG 5.2

The cost of renewable energy is falling.

Source: The Economist, 2023.

***

In South Baltimore, a group of teen student activists in Free Your Voice successfully challenged the giant freight transportation company CSX, which transported more than 8 million tons of coal through the area in 2021. Shashawnda and her fellow students went door to door, warning their neighbors about the dangers of the project. The city school system was planning to purchase electricity from the incinerator. So, the students organized and took their opposition to the school board. They even wrote protest poetry and music challenging a very different kind of danger in their neighborhood: air pollution and climate change.

Following is a sample of the poetry they shared:

If the incinerator takes away a breath;
How many do we need until there is nothing that's left?
Until the smoke clogs up and we can't feel our chest.

They received a standing ovation from the board—and convinced the school system to pull out of their contract. Other agencies that had lined up to purchase energy from the project also pulled out, and the company didn't build the incinerator.

Source: Parker et al., 2023.

***

A group of young people from Montana won a major climate case on Monday after arguing the state failed to protect their right to a clean environment by continuing to use fossil fuels.

The ruling determined that a provision in Montana's Environmental Policy Act violated the right to a clean environment, which is guaranteed under Montana's state constitution, by promoting the continued use of fossil fuels.

The court said a provision in the law that prevented the state from considering the climate impacts of energy projects is unconstitutional.

Source: Ebbs & Grant, 2023.

The Ecobiopsychosocialspiritual Aspects of Faith

In this last subsection, we will briefly examine the sources and characteristics of faith. As we noted earlier, faith is closely interrelated with the many elements of our ecobiopsychosocialspiritual nature. As Arthur Dobrin (2012) put it, "Faith speaks the language of the heart. It is an expression of hope that goes beyond the conscious mind" (p. 1). If faith is based in the heart, then does this mean that faith does not also have biological, social, and cognitive elements?

I think it is helpful for professional helpers to see faith as a language of the heart. It is equally helpful to think about how we can inform our heart-based faith by also drawing from the wisdom of our body, mind, *and* spirit. We can use the logic of our minds, for example, to help us assess whether we have realistic expectations. We may err in either direction—either expecting a rapid and significant change that is unlikely or underestimating the extent to which change is possible. Both kinds of errors are common, and there is evidence that we especially tend to underestimate our own ability to make a difference. Social scientists now believe that civilizations are actually *more* able to change than most think they are (*New Scientist*, 2023).

Our spirituality and religiosity can support the faith that inspires us to become involved in change. There is a history of spiritual and religious leaders who led through their faith, and today, an increasing number of clergy believe that they have a responsibility to become activists and lead their congregations in addressing such challenges as the climate crisis (Randall, 2023).

In the United States, various religions and cultures have had different relationships with the climate crisis. Between 2014 and 2023, the percentage of people in the United States who saw climate change as a crisis increased in those identified with Judaism or Hispanic Catholicism, whereas the percentage decreased with those identified with White evangelical Protestant beliefs. Across the same time period and across all Americans in the United States, both affiliated and unaffiliated with religion, the percentage of people who saw climate change as a crisis increased overall from 23% to 27%. Finally, the percentage for all people not affiliated with religion who saw climate change as a crisis increased from 33% to 43%. Professional helpers, as we will discuss more in Sections 2 and 3, can bring different religious or culture groups together to participate in public intergroup dialogues, which can lead to increases in mutual understanding, cooperation, and, eventually, the discovery of common across differences (Derezotes, 2014).

We also, of course, exist in our body, which also can contribute to faith. Bodies have wisdom. I find, for example, that my body will tell me, if I listen, when to rest and when to work. Professional helpers can model faith-based activism for our communities that is rooted in such body wisdom, in which we find a balance of work and play, rest and effort. Our body-knowing can help us confirm what we think we know through other means. For example, people often use such expressions as "I had a gut feeling" or "My heart told me." Such below-the-neck experiences can help confirm other ways of knowing. We will further explore this kind of work and connection with our own bodies in future chapters.

QUESTIONS FOR REFLECTION

1. Has your reaction to the climate crisis tended to be more of hope or despair? Has this changed over time? Please explain.
2. After reading this chapter, have your views about hope, despair, and faith changed at all? Please explain.

3. This chapter begins with the words of Estes: "My friends, do not lose heart. We were made for these times." What is your reaction to these words? If one believes in this message, how does that change one's willingness and ability to be an activist?

4. Do you see yourself as an *activist*? What does that word mean to you? Please explain.

REFERENCES

Blanch, A. (2018). *Power vs. empowerment—What's the difference and why is it important?* Good Psychology, June 17, 2018. https://www.good-psychology.com/post/power-vs-empowerment-what-s-the-difference-and-why-is-it-important

Brei, A. T. (Ed.; 2016). *Ecology, ethics and hope*. Rowman & Littlefield.

Carter, J. (1995). *Keeping faith: memoirs of a president*. University of Arkansas Press.

CBS Mornings. (2023). Jimmy *Carter's pastor discusses former president's legacy, faith and lifetime of service*. YouTube, February 21, 2023. https://www.youtube.com/watch?v=o9KfqHc86zc

Celebrity Beliefs. (2017). *Michael Jackson*. October 8, 2017. http://www.celebritybeliefs.com/michael-jackson/

Derezotes, D. S. (2014). *Transforming historical trauma through dialogue*. Sage Publications.

Dobrin, A. (2012). Why faith is important: Faith speaks the language of the heart. *Psychology Today*, September 28, 2012. https://www.psychologytoday.com/us/blog/am-i-right/201209/why-faith-is-important

Ebbs, S., & Grant, T. (2023). *Montana youths win climate lawsuit against state for promoting fossil fuels*. ABC News, August 14, 2023. https://abcnews.go.com/US/montana-youths-win-climate-lawsui -fossil-fuels/story?id=102260674

The Economist. (2023). Renewable energy has hidden costs. September 21, 2023. https://www.economist.com/finance-and-economics/2023/09/21/renewable-energy-has-hidden-costs

Estes, C. P. (2023). *We were made for these times*. https://www.awakin.org/v2/read/view.php?tid=2195

Fraser, J. S. (2019). Hope: A foundation of all psychotherapy that works. *Psychology Today*, June 8, 2019. https://www.psychologytoday.com/us/blog/breaking-the-cycle/201906/hope-foundation-all-psychotherapy-works

Glauser, M. (2022). *One people one planet: Six universal truths for being happy together*. Lioncrest.

Goodreads. (2023). *Albert Einstein quotes*. https://www.goodreads.com/quotes/429690-the-most-important-decision-we-make-is-whether-we-believe

Green, R. M. (Ed.; 2018). *Theories of hope: Exploring alternative affective dimensions of human experience*. Rowman & Littlefield.

Higa, B. (2023). Hope for the hopeless. *Buddhadharma*. Buddha-Nature. https://buddhanature.tsadra.org/index.php/Articles/Hope_for_the_Hopeless_(Higa_2023)

Johnson, R. (2018). *Inner gold: Understanding psychological projection*. Koa Books.

Kelsey, E. (2020). *Hope matters: Why changing the way we think is critical to solving the environmental crisis*. Greystone.

Keniston, K. (1968). *Young radicals: Notes on committed youth*. Harcourt Brace & World.

Krishnamurti, J. (1996). *Total freedom: The essential Krishnamurti*. Krishnamurti Foundation of America.

Krishnamurti, J. (2022). *Krishnamurti on hope*. https://krishnamurti.podbean.com/e/krishnamurti-on-hope/

Krishnamurti, J. (2023). *Series II: Chapter 21: Despair and hope*. JKrishnamurti.org. https://www.jkrishnamurti.org/content/series-ii-chapter-21-despair-and-hope

Lear, J. (2006). *Radical hope: Ethics in the face of cultural devastation*. Harvard University Press.

Macy, J. (1983). *Despair and personal power in the nuclear age*. New Society Publishers.

Macy, J. (2023a). *Joanna Macy and her work*. https://www.joannamacy.net/main

Macy, J. (2023b). *Active hope information*. https://www.activehope.info/

Macy, J., & Johnstone, C. (2012). *Active hope: How to face the mess we're in without going crazy*. New World Library.

New Scientist. (2023). Change is possible: The nature of civilization is far more malleable than we thought. July 1, p. 5.

Obama, B. (2014). *Choosing hope: President Obama's address to the United Nations*. https://obamawhitehouse.archives.gov/blog/2014/09/24/choosing-hope-president-obama-s-address-united-nations

Parker, B. A., Hersher, R., Stein, C., Qureshi, B., Mortada, D., Banerjee, N., & Retting, A. (2023). *Student activists are pushing back against big polluters—and winning.* National Public Radio, October 4, 2023. https://www.npr.org/2023/10/04/1197954102/student-activists-climate-change-baltimore

Randall, R. (2023). Spirituality, global warming, and grief: How clergy can help tackle climate anxiety. *Mother Jones*, June 18, 2023. https://www.motherjones.com/environment/2023/06/spirituality-global-warming-and-grief-how-clergy-can-help-tackle-climate-anxiety/

Reader's Digest. (2023). *55 powerful Martin Luther King, Jr., quotes that stand the test of time.* https://www.rd.com/list/the-best-quotes-of-martin-luther-king-jr/

Schaffner, A. (2020). *Perseverance in psychology: Meaning, importance, and books.* PositivePsychology.com, September 16, 2020. https://positivepsychology.com/perseverance/

Seligmann, M. E. P. (2019). *The hope circuit: A psychologist's journey from helplessness to optimism.* Public Affairs.

The Spiritual Life. (2023). *Martin Luther King, Jr.'s, quotes.* https://slife.org/martin-luther-king-jr-s-quotes/

Williamson, M. (2023). *Marianne Williamson: "The Spirit of America" 09/03/2023.* YouTube, September 11, 2023. https://www.youtube.com/watch?v=WRVajnyavOw

Yalom, I., & Leszcz, M. (2020). *The theory and practice of group psychotherapy* (6th ed.). Basic Books.

Yeshe, A., Grace, C., & Sofer, O. J. (2023). Ask the teachers: What is the Buddhist view of hope? *Lion's Roar*, May 20, 2023. https://www.lionsroar.com/ask-the-teachers-what-is- the-buddhist-view-of-hope/

Credits

SECTION 2

Climate Change in Humansystems

IMG 1.1

IMG 1.2

As we said in Section 1, the main thesis of this text is that the roots of our climate crisis can be found in our own human consciousness, and as we change our consciousness, we can help heal and transform the climate of our interrelated human systems and ecosystems.

What is a *humansystem*? Humansystems are the complex webs of human connections and interactions within and between such groupings as the family, work setting, institutions, and local and global communities. In the two images shown here, we see people in an urban and in a rural area of the United States, which represent two example of humansystems in our country.

A climate change in a humansystem occurs when we change our behaviors within or between such humansystems. For example, members of a family might decide to put solar panels on their roof. Employees might decide to strike until their demands for their corporation to stop emitting greenhouse gases are met. Or perhaps the people in a city elect a new mayor who has promised to support cleaner air and more green energy projects.

In Section 1, which includes Chapters 1 through 5, we looked at how the state of our ***human consciousness*** is associated with our climate crisis and with potentially with desired climate change in humansystems and ecosystems. We showed how increased *consciousness*, which we defined as increased reverent awareness, can help humanity heal and transform ourselves and our world.

In Section 2, we will show how ***dialogue practice*** can help us heal and transform humansystems so that humanity can better cooperate together to heal the ecosystems that support all life on Earth. We will examine inclusive dialogic strategies of change that professional helpers can use when working with climate crisis, at the micro, mezzo, and macro levels of human systems. Although some theory and some interventions are provided in every chapter, we will see more practical interventions in the chapters in Sections 2 and 3.

The first chapter in Section 2, Chapter 6, provides a generalist introduction, in which we study the core inclusive practice engagement, assessment, and intervention strategies for work with humansystems, with an emphasis on dialogue practice. In Chapter 7, we study the polarization that appears to be a pandemic in our local and global communities today and intervention strategies that can lead to healing and transformation. Climate justice is the topic of Chapter 8, in which we look at issues of diversity, equity, and inclusion as well as intervention strategies. In the last two chapters, we introduce climate-sensitive practice, in Chapter 9 with families and institutions and in Chapter 10 with local and global communities.

Credits

CHAPTER 6

IMG 6.1

Inclusive Practice With Humansystems

Human life and humanity come into being in genuine encounters. The hope for this hour depends upon the renewal of the immediacy of dialogue among human beings.

—MARTIN BUBER (INSPIRING QUOTES, 2024)

When two people relate to each other authentically and humanly, God is the electricity that surges between them.

—MARTIN BUBER (QUOTLR, 2024)

Years ago, I was asked to facilitate a lunchtime dialogue for participants attending a conference on water. I had only about 50 minutes. There was not enough time to dive very deep into a challenging dialogue.

Attending the conference was a diverse mix of people who all had an intense interest in the use of water in our part of the American West, including ranchers, politicians, members of local Indigenous tribes, local and federal government administrators, university academics and their students, environmentalists, and other advocates for both rural and urban areas.

The morning sessions had been tense, with often heated arguments between participants following the breakout presentations. People filed into the lunch area to pick up their box lunches, preferring to sit at tables with others who seemed to think and look like them.

I asked the conference coordinator to please not say much when she introduced me. I wore the jeans and a short-sleeved camp shirt that I like, anticipating that no matter what I wore, different people would have different reactions to me. It was probably best to feel comfortable.

There were maybe 60 people in the room, a little large for a whole-group discussion. If we did breakout groups, however, that exercise in itself might take more than half of the time we had together. I decided to set up a whole-group conversation.

Grabbing the handheld wireless stick-microphone I had requested, I walked up to a random table and welcomed the people there. I asked one gentleman how he was today, and after he replied into the mic, I then asked him what his favorite lunch restaurant is, anywhere in the world, and why. After he responded, I walked over to another table and repeated the same questions with another person. People were beginning to listen across the room now, peering at us over their box lunches. Some were smiling.

After talking with a few more people, I turned to the audience and asked to them to remember a story about water from their childhood that they would be willing to share. People had not expected this kind of conversation, of course, and there was immediately some laughter and talk at the tables.

I asked for a volunteer to start, and immediately, a few hands went up. A man talked about playing with his brothers and sisters with giant squirt guns, spraying each other on a hot summer day. A woman volunteered a story about the time she leapt of a cliff into a river near her house, after her friends dared her. Several people had stories about going fishing in their favorite lakes or streams. An older woman talked about her adventures finding and exploring water holes in the desert with her husband, and discovering tiny fish and insects that inhabited these precious spots. There were other stories about water balloons, waterfalls, rainstorms, floods, and rainbows. There was laughter and delight in the room. And a little light.

After a morning of intense debate, people were starting to notice each other's humanity again.

By this time, we had about 15 minutes left. I asked them why they thought I had asked them to do this kind of sharing. They understood. One woman explained that she had felt frustrated all morning because many people had unfortunately not been very civil with each other and no common ground had been identified. Others agreed.

A conversation began in the last minutes about the importance of civility and trust. I was pleased that people had noticed how the group process had been going and had wanted to relate differently with each other.

The afternoon sessions were different. Not perfect, not easy, but people were noticeably more civil. And they sat with different people at the tables during the afternoon break.

What Is Dialogue Practice?

The purpose of dialogue practice is to support the development of human relationships that bridge the differences that currently divide us. In this chapter, we will explore the use of dialogic inclusive practice principles in fostering climate change in our humansystems. In Chapter 1, we introduced some principles of inclusive practice and applied them to our work with human consciousness. In this chapter, we will apply those principles to our dialogue practice with human systems.

What is dialogue practice? We will suggest that most of what we call professional helping with human-systems is a form of dialogue practice. Maurice Friedman (1983) described the dialogue practice of Martin Buber's as a "confirmation of otherness" in which people strive to understand and accept each other, regardless of our differences.

From my perspective, we professional helpers often work as dialogue facilitators, as we help people develop consciousness of themselves and of each other. Whether we are doing individual mentoring, couple and family therapy, teaching small classes in a college, or facilitating intergroup dialogue in community settings, we are doing dialogue facilitation.

We Are a Lonely Country Divided by Differences

When I was in high school, back in the 1960s, we all sat at our lunch tables everyday with people who looked and acted like us, similar to how the adults sat together in my earlier story about the water conference. The tables were segregated by race, gender, social economic status, and cultural identity. Sadly, today, 60 years later, it appears that high school students and we adult professionals for that matter, still often sit in similar segregated groups at lunch time. This kind of behavior seems to be both partly learned and partly rooted somehow in human nature. We want to avoid discomfort and feel safe and secure.

Unfortunately, few people I know today seem to feel comfortable, safe, and secure. As we discussed in earlier chapters, many young people in our country instead report feeling isolated and lonely. Teens report anxiety and depression are major problems and many feel that their support network could be improved; even higher proportions of low-income populations and many people of color often report similar loneliness and isolation (Pew Research Center, 2020). Overall, about one-third of adults report loneliness, and 58% of the U.S population say that they often feel no one knows them well. Increasing numbers of people report that they are seeking more friendships and connectedness than ever, a trend that is especially true for women (Rameer, 2023).

How did we get here? What's going on? At the same time that our younger generations have access to greater communication technology than any humans throughout history, they also seem to be experiencing significant disconnection from other people, as well as from outdoor natural surroundings, with likely negative impact upon their overall well-being (Scott et al., 2022).

It appears that many of our private and public conversations today are monologues, rather than dialogues. In a monologue, only one person is speaking, and no one is listening. Monologue is often created when there is not enough safety. There may be a subtle or overt threat of violence, psychological or physical, that silences the Other. In contrast, in dialogue everyone gets to speak and everyone listens.

Dialogue practice does not eliminate the differences between human beings, nor do we want to try to eliminate them. Those differences are beautiful and create variety in our lives. Our genetic differences also support community well-being; we know that evolution favors subtle variants that give human populations the ability to survive and prosper in their own unique environments (Smithsonian, 2023).

Inclusive Practice and Dialogue Practice

Let's review the key elements of inclusive practice, which, as we explained in Chapter 1, informs the climate-sensitive practice theory we are developing in this text. In Table 6.1, we list the elements, beginning with consciousness, that we have defined in our text as reverent awareness of self and the world. As we have theorized in our text, the climate change we want to see in humansystems and ecosystems is based on a climate change in human consciousness.

We also emphasize the use of all ways of knowing, which include science, reason, intuition, and imagination (McGilchrist, 2009). None of the four ways of knowing is favored over the others, and all four ways are utilized, when possible, in every practice situation. When data generated by several, and ideally all four, ways of knowing suggest the same conclusion, we can have increased confidence that we understand what is going on.

Both practice-based evidence and evidence-based practice approaches are used in our inclusive assessments. Practice-based evidence comes from our own assessment of our practice, whereas evidence-based practice is typically based on research findings from refereed journal articles (see Duncan et al., 2010).

TABLE 6.1 Key Elements of Inclusive Practice

Elements	Brief Descriptions
First of all, consciousness	We strive to foster reverent awareness of self and the world.
Use all ways of knowing	We use all four ways of knowing: science, reason, intuition, and imagination.
Use the art and science of practice	Professional helpers serve as scholar-practitioners.
Promote inclusive communities	We strive to transform humansystems and ecosystems into inclusive communities that are welcoming, just, equitable, and diverse.
Use person-in-environment, contextual approach	We understand and address the micro-, mezzo-, macro-, and eco-level contexts.
Use evidence-based practice	We research scientific methodologies and findings.
Engage	Professional helpers engage with genuineness and loving kindness.
Assess ecobiopsychosocialspiritual elements	We want to foster multidimensional lifelong development.
Use all practice paradigms	We use case management, psychodynamic, cognitive-behavioral, experiential, transpersonal, body-mind, and deep ecology.
Use dialogic practice orientation	We strive to foster relationships through confirmation of otherness and bridge the differences that divide us.
Use practice-based evidence	We assess the progress in our own practice.

Inclusive practice understands humans from a person-in-environment, or contextual, perspective. This means that we examine the complex interconnections between the micro, mezzo, macro, and eco levels of each client's life situation.

Inclusive practice promotes the transformation of humansystems into inclusive communities. What does that mean? Inclusive communities welcome all members, voices, cultures, and perspectives. As we have discussed, dialogic communication is used in inclusive communities, a topic that we will examine further in this chapter.

Our assessments and interventions address all the dimensions of human development. We use the term *ecobiopsychosocialspiritual* to describe this inclusive approach. In Table 1.1 in Chapter 1, we offered examples of the five ecobiopsychosocialspiritual dimensions of assessment in inclusive practice. In Table 6.2, we offer examples of how we might address each of these dimensions in our dialogue facilitations on the climate crisis.

TABLE 6.2 Ecobiopsychosocialspiritual Questions for Dialogue Practice

Dimension	Basic Focus	Examples of Questions for Dialogue
Eco (ecological)	Our relationship with our human-made and "natural" environments	What is your culture's creation myth? How does nature nourish you? How do you feel toward other living things?
Bio (biological)	Our relationship with our bodies	How were you taught about your body? What is your relationship with your body today? How do you relate with other bodies?
Psycho (psychological)	Our relationship with our emotions and thoughts	What is my relationship with my own mind? How sensitive am I to the feelings of others? Can I express my emotions and feelings?
Social	Our relationship with other people	What kind of relationships do I have with others? What kinds of relationships do I want and need? How do I relate with other living things?
Spiritual	Our relationship with our own spirituality	Do you have a sacred landscape? Do you have a "power animal" or "power plant"? How and when do you feel spirituality?

In our interventions, inclusive practitioners draws from all practice paradigms. We recognize that all these concepts can have utility in certain situations and seek to match the best interventions with the data we have collected in our inclusive assessments. In many situations, multiple methods of counseling, teaching, or facilitation may be useful to our clients, students, and participants. Many examples of interventions that we can use with humansystems and ecosystems are provided in Sections 2 and 3.

In Table 1.3 in Chapter 1, we presented a brief summary of the seven paradigms of social work practice along with the basic goals of each paradigm, the developmental dimensions they especially address, and examples of some models that fit in each category. In Table 6.3, we provide some examples of interventions that could be used in working with humansystems when addressing the climate crisis.

TABLE 6.3 Examples of Interventions With Humansystems Drawn From Seven Paradigms of Practice

Paradigms	Examples of Interventions
Psychodynamic	Ask participants in a community dialogue to talk about how they each experienced the trauma of having part of their town burn down in a wildfire several years ago.
Cognitive-behavioral	Challenge middle school students to share in class the beliefs they have about the value of having city parks and local forests.
Experiential-phenomenological	Bring a group of retired people together to create and manage an urban garden.
Transpersonal	Ask participants in a seminar to talk about their own "sacred landscapes."
Case management	Help organize people living in a dense urban area create pedestrian and bike paths in their neighborhoods.
Biopsychosocial	Organize outdoor yoga and tai chi classes in a park.
Ecobiopsychosocial (or deep ecology)	Bring therapy animals into our work with clients who are in an inpatient mental health program.

Source: Derezotes, 2023.

Dialogue Facilitation

Like all living things, we human beings have a natural connection with the universe in which we live. The dialogic process can help us experience and develop these connections as we make sustainable and cooperative relationships with other people, other living things, and the ecosystems that support life. Dialogue practice challenges participants to own and accept our own emotions, thoughts, feelings, and beliefs. We learn how to form relationships across the differences that currently divide us and to listen to each other without reacting with anger or defensiveness.

The dialogue facilitator's intent is to help co-create an inclusive community that supports both connection between members as well as the diversity seen in any human population. *Dialogue* can be defined as:

> a nonviolent relationship-building practice, with shared commitment to participate in intrapersonal, interpersonal, and ecological work, directed towards increasing the mutual understanding, acceptance, and empowerment necessary for our collective transformation from "trauma-repetition" towards sustainable, equitable, and peaceful communities. (Derezotes, 2014, p. 44)

Although interest in dialogue continues to grow, the practice has not yet been widely implemented in our educational systems. Despite the need for dialogue in our contentious world, we still do not teach or model dialogue to most children or young adults. Few high schools offer opportunities to participate in dialogues, although most have debate teams, and dialogue is typically not yet required in most higher education professional programs.

Researchers have found that dialogue practice actually can help people form relationships across the differences that divide us. Dialogue can help people move beyond our tendency to demonize and lose empathy for other populations with whom we are in conflict (Hedges, 2018). Dialogue participants learn skills in listening and responding as well as mutual empathy, authenticity, mutuality, and sensitive humor that can help transform relationships (Hartling et al., 2000).

A literature review found that higher education dialogue programs offer approaches that enhance communication, trust, and relationship and use a combination of intellectual discussion and emotional sharing. In addition, students engaged in campus dialogues report enhanced understanding of other cultures, increased problem-solving skills, and a greater awareness of social inequality and of group identities (Dessel & Rogge, 2008). Dialogue, when combined with a follow-up service activity, can be especially effective in promoting personal value development and increased civic responsibility (Sutton, 2010).

In his research on dialogue outcomes, Nagda (2006) found that participants learned four key dialogic processes, including critical self-reflection, engaging the self, understanding differences, and making alliances. Dialogue participants were found to learn in four key areas: knowledge about the out-group, new ways of thinking and acting, developing emotional connections, and challenging stereotypes (Pettigrew, 1998). In general, both oppressed and privileged groups report benefits in relationship building and mutual understanding from dialogue participation (Dessel & Rogge, 2008).

Dialogue has been found to have benefits to participants across the world. For example, participants in intergroup dialogues in Peru, Argentina, Guatemala, Panama, and the Philippines reported more reduced stereotypes, greater mutual understanding, and increased support of social justice policy change. Positive relationship-building outcomes from dialogue have been found in many international conflict resolution efforts (Rieker & Thune, 2015).

There are perhaps two central challenges in dialogue facilitation. The first is to get people to come to the dialogue and open up and get real, so to speak, and the second is what to do when they *do* open up.

Dialogue facilitation is challenging but rewarding work. The most effective facilitator has a number of characteristics. The facilitator characteristics that predict therapeutic alliance include genuineness, empathy, alliance, cohesion, therapist disclosure, and listening (Norcross, 2010). We also know that in professional helping, the rationale for the treatment and the approach that is based on the rationale are most effective when both client and helper both believe in rationale and approach and agree on the goals and tasks within their therapeutic alliance (Wampold, 2010).

The dialogue facilitator is willing to self-reflect and go inside themselves to recognize how they really feel and think. The facilitator is also a genuine person who is able to show people they care about them and is also able to challenge them when necessary. Successful professional helpers also have the ability to develop effective helping relationships (Norcross, 2010). Since dialogue is a relationship-building process, the effective dialogue facilitator is skilled in relationship building in a group process.

Dialogue Ground Rules

We say that without civility, we can have no civilization. Civility is not the habitual and often unconscious "niceness" that many people display in public settings to either please other people or protect themselves when they are afraid. We instead can think of civility as a conscious choice to offer loving kindness to others, regardless of what we may think or feel about them (see Boatright et al., 2019).

In Table 6.4, we offer a list of 14 ground rules for dialogue that can assist facilitators and participants in acting civilly with each other in dialogue meetings. Dialogue facilitators can either offer the list before the conversation begins or ask participants to create their own list, which usually ends up containing very similar items.

Most of these items are self-explanatory. Probably the two most important are to listen for understanding and to speak respectfully. Many of us do not listen for understanding when others are speaking but rather are thinking about what we want to say when they are finally done. *Respect* is a little hard to define, but we can say that the golden rule applies here—respect is the attitude we want to receive from others that we freely offer to them first. The offering of amnesty is about forgiving others when they make mistakes, because we all make mistakes. Number 7 under the *DO* list refers to Mindell's (1995) idea that we can chose to share what our biases, judgments, and projections in a respectful way, as a means of healing ourselves and establishing trust in a group. Number 7 under the *DON'T* list is the idea that we refrain from talking from an authoritarian or superior position to others and acting like the infallible expert.

TABLE 6.4 Dialogue Ground Rules

DO	(1) Listen for understanding
	(2) Speak respectfully
	(3) Speak for myself
	(4) Offer amnesty
	(5) Allow at least three or four people to talk before speaking again
	(6) Allow others to have different experiences, beliefs, and feelings
	(7) Own my own biases, judgments, and projections
DON'T	(1) Interrupt
	(2) Make others wrong
	(3) Engage in side talk
	(4) Make disrespectful gestures
	(5) Talk for longer than a few minutes during my turn
	(6) Do side talk
	(7) Do teaching and preaching

Some Basic Dialogue Tasks and Methods

Often one of the most difficult tasks is to find participants who are willing to enter a dialogue and commit to participating. People can be understandably nervous about meeting with other people who may hold different beliefs, and usually most are also unsure about what a dialogue actually is. Sometimes I have asked participants to commit to a series of dialogues, if possible, because it typically takes time for participants to become ready to share how they may feel and think. Sometimes we use the model of having a brief informational presentation before the dialogue begins, in which the facts about an issue are presented so that the participants are all starting with the same information. Even this approach can be controversial in these days of climate change deniers, so the presenter might only offer the most well-established facts and not less established opinions. This is especially true in a situation in which a controversial topic is being tackled in an intergroup dialogue; the facilitator wants to try to find approximately equal numbers of attending people on each side of the issue.

Facilitators help create the dialogic space by modeling the ground rules themselves and by facilitating the initial introductions and check-in, when people usually are asked to say hello and respond to warmup questions. Such questions can be fun (e.g., what is your favorite breakfast?), deeper (e.g., what are your hopes and fears about this group?), or both. Facilitators constantly strive to find the balance between fostering more group safety (although there is no such thing as total safety, we can reduce the fears that participants typically have) and challenging participants to engage in deeper and more meaningful conversations. When difficult conversations do begin, the facilitator helps participants "sit in the fire" (Mindell, 1995), which means that instead of running away from our discomfort, we continue to participate and follow the ground rules as best we can.

How much structure should the facilitator provide? In general, the less mature and more fearful or reactive the participants are, the more structure needs to be provided. The facilitator might reduce fear and anger and increase civil participation by engaging people at the beginning of the dialogue in an activity, such as asking people to respond to a specific question or two.

In Table 6.5, we offer a list of phases and tasks in the dialogue process. This is just a guideline, because every dialogue will look at least a little different, depending on the context and participants.

TABLE 6.5 Phases and Potential Tasks in Dialogue

Phases	Tasks
Engagement and assessment	Identify needs
	Identify participants
	Set goals
	Get commitment from participants and stakeholders
	Establish ground rules
Relationship-building work	Create the dialogic space
	Participants check in
	Engage in difficult conversations as appropriate
	Engage in dialogue activities and then briefly process what happened
	Rotate between whole-group and breakout-group dialogue activities
	Participants check out
Evaluation and follow-up	Negotiate additional meetings if appropriate
	Refer participants to other programs as needed
	Measure outcomes
	Offer opportunities for participants to engage in social action

Source: Derezotes, 2014.

Dialogue Models

The dialogue facilitator has a number of models they can use to inform their facilitation. Dialogue models can be organized into seven categories by applying the seven paradigms of inclusive practice we introduced in the text. The dialogue models thus include the psychodynamic, cognitive behavioral, experiential, transpersonal, whole-body, and deep ecology approaches (Derezotes, 2014).

Psychodynamic dialogue is about the past and usually involves people sharing their stories. We can say that the shortest distance between two people is often bridged by a story, and storytelling can help us start to respond with increased warmth, empathy, and compassion toward each other. For example, in an intergroup dialogue I facilitated on campus between left- and right-leaning students, it was fascinating to discover the stories behind why individual students became Republicans or Democrats.

Another way to connect with each other is by owning our beliefs and behavior patterns. In cognitive-behavioral dialogue, the conversation is about how we think and act. Since most of us tend to be in our heads, these dimensions are often easier for people to discuss first, in contrast with dimensions such as emotions and spirituality. However, even in cognitive-behavioral dialogue, since people today are often strongly identified with their beliefs, such as about politics, religion, or culture, we may also observe defensive or angry reactions when using this model.

In experiential dialogue, we ask participants to share their emotions with each other. Many professional helpers know how to teach their clients how to use "I" messages (e.g., "I feel sad, mad, glad, scared, excited") rather than "you" messages (e.g., "You are so annoying"). We also can teach people how to share emotions in an assertive manner (e.g., "I notice that I feel very sad about what you told me") as opposed to an aggressive (e.g., "I hate you and hope you die") or passive (e.g., the person gets silent and pouts during the dialogue) manner.

Transpersonal dialogue adds the spiritual dimension into the conversation. Although most of us report having spiritual experiences, we often keep them very private. Spiritual sharing may include dialogue about one's sense of meaning or purpose in life; identification of and appreciation of what one finds sacred; and experiences of connectedness, peace, and forgiveness. Many people report that they experience spirituality most frequently in certain life experiences, such as a walk in nature or witnessing the birth of a child.

We humans often communicate with each other through nonverbal physical movements and expressions (Murphy, 1993). Biopsychosocial, or *whole-body*, dialogue uses this ability to communicate nonverbally. I was first introduced to nonverbal group exercises many years ago by a dance instructor who asked us to each create a 15-second-long dance movement that we then could use in front of the group. This introduction activity was freeing for me and an effective and enjoyable way to connect initially with others. Another nonverbal intervention is to have people draw a picture of themselves that they later share with others.

Dialogue activities in deep ecology can help people become more aware of their connections with other living things and their ecosystems. Deep ecologists are especially interested in the interconnectedness and equal value of all living things (Tobias, 1988), and, indeed, we continue to find evidence that human well-being is interconnected with the well-being of other living things and our ecosystems (de Steiguer, 2006). One of my favorite deep ecology dialogue techniques is to have some participants role play other animals or plants in a conversation in which other participants play the role of humans.

This basic introduction to dialogue practice provides a foundation for inclusive practice with humansystems and ecosystems. We will be offering examples of dialogue approaches throughout the rest of the text that can be used in various contexts to address our climate crisis.

QUESTIONS FOR REFLECTION

1. Have you ever participated in a debate? What was that experience like?
2. Have you ever participated in a dialogue? What was that experience like? How was it different from the debate (if you had experienced both)?
3. Would you like to learn dialogue facilitation skills and then use them professionally? Why or why not?

REFERENCES

Boatright, R. G., Shaffer, T. J., Sobierai, S., & Young, G. (Eds.; 2019). *A crisis of civility? Political discourse and its discontents.* Routledge.

de Steiguer, J. E. (2006). *The origins of modern environmental thought.* University of Arizona Press.

Derezotes, D. S. (2014). *Transforming historical trauma through dialogue.* Sage Publications.

Derezotes, D. S. (2023). *Inclusive social work: A new vision of community practice.* Cognella.

Dessel, A., & Rogge, M. E. (2008). Evaluation of intergroup dialogue: A review of the empirical literature. *Conflict Resolution Quarterly, 26*(2), 199–238. https://doi.org/10.1002/crq.230

Friedman, M. S. (1983). *The confirmation of otherness, in family, community, and society.* Pilgrim Press.

Hartling, L., Rosen, W., Walker, M., & Jordan, J. (2000). Shame and humiliation: From isolation to relational transformation. *Work in Progress, 88*, 1–14. Stone Center for Developmental Services and Studies. https://www.humiliationstudies.org/documents/hartling/HartlingShameHumiliation.pdf

Hedges, C. (2018). *America: The farewell tour.* Simon and Schuster.

Inspiring Quotes. (2024). *Martin Buber famous quotes.* July 22, 2024. https://www.inspiringquotes.us/author/4790-martin-buber/about-dialogue

McGilchrist, I. (2009). *The master and his emissary: The divided brain and the making of the Western* world. Yale University Press.

Mindell, A. (1995). *Sitting in the fire: Large group transformation using conflict and diversity.* Lao Tse Press.

Murphy, M. (1993). *The future of the body: Explorations into the further evolution of human nature.* Tarcher.

Nagda, B. A. (2006). Breaking barriers, crossing borders, building bridges: Communication processes in intergroup dialogues. *Journal of Social Issues, 62*(3), 553–576.

Norcross, J. C. (2010). Chapter 4: The therapeutic relationship. In Duncon, B. L., Miller, S. D., Wampold, B. E., & Hubble, M. A. (Eds.), *The heart and soul of change: Delivering what works in therapy* (2nd ed.; pp. 113–142). American Psychological Association.

Pettigrew, T. F. (1998). Intergroup contact theory. *Annual Review of Psychology, 49*, 65–85.

Pew Research Center. (2020). *How Americans feel about the satisfactions and stresses of modern life.* February 5, 2020. https://www.pewresearch.org/short-reads/2020/02/05/how-americans-feel-about-the-satisfactions-and-stresses-of-modern-life/

Quotlr.com. (2024). *110+ Martin Buber quotes on relationships, dialogue and existentialism.* April 26, 2024. https://quotlr.com/author/martin-buber

Rameer, V. M. (2023). *U.S. loneliness statistics 2023: Are Americans lonely?* Science of People, January 16, 2024. https://www.scienceofpeople.com/loneliness-statistics/

Rieker, P., & Thune, H. (2015). *Dialogue and conflict resolution: Potential and limits.* Routledge.

Scott, S., Gray, T., Charlton, J., & Millard, J. (2022). The impact of time spent in natural outdoor spaces on children's language, communication and social skills: A systematic review protocol. *International Journal of Environmental Resources and Public Health, 19*(19), 12038. https://doi.org/10.3390%2Fijerph191912038

Smithsonian. (2023). *What does it mean to be human?* https://humanorigins.si.edu/evidence/genetics

Sutton, S. B. (2010). Experiencing dialogue: Conversation as transformation. *Practicing Anthropology, 32*(3), 49–53.

Tobias, M. (Ed.; 1988). *Deep ecology*. Avant Books.

Wampold, B. E. (2010). Chapter 2: The research evidence for common factors models: A historically situated perspective. In Duncon, B. L., Miller, S. D., Wampold, B. E., & Hubble, M. A. (Eds.), *The heart and soul of change: Delivering what works in therapy* (2nd ed.; pp. 49–82). American Psychological Association.

Credit

CHAPTER 7

IMG 7.1

Polarization and the Radical Middle

Rather than being defined by consistently applied principles—about the right to a democratically controlled public square, say, and to trustworthy information and privacy—we have two warring camps defining themselves in opposition to whatever the other is saying and doing at any given time … [although these] camps are not morally equivalent.

—NAOMI KLEIN (2023)

The radical middle says it's time to take our politics back from this "he said, she said" mentality. Most Americans aren't nearly as polarized as the Presidential campaign suggests. Most of us aren't at some mushy middle point, either. Most of us want to see a new kind of politics, where we take the genuine and often very reasonable concerns of Democrats and Republicans, and build on them toward something different, something exciting and new. … Those of us who've adopted it are daring to suggest real solutions to our biggest problems, but we're doing so from the middle. We're listening to radicals and liberals, we're listening to conservatives and libertarians, and we're not losing touch with the often complicated facts on the ground.

—MARK STATIN (*RADICAL MIDDLE NEWSLETTER*, N.D.)

After the September 11, 2001, attacks, I was fortunate to be invited to join a wonderful group of clinical psychologists with whom I am still close friends. We wanted to somehow address the social-political divisions that seemed at the root of those terror events. We were interested in how it happens that people become willing to dehumanize and murder each other, and in how to change that.

At about the same time, I was also fortunate to be invited to join and later direct the Peace and Conflict Studies program at my university. I was always curious about how the differences that divide us globally mirror the same differences that divide us locally and could see how those tendencies toward dehumanization and violence are in everyone, including myself.

In the university, for example, faculty might eloquently lament in their classrooms about the polarization in our government today. Then, soon afterward, the professors might stop in the hallway to share the latest gossip about another colleague before hurrying to college council where we can bitterly engage in the latest departmental power struggles.

Alternatively, a history book might reveal how some unhappy U.S. civil war generals sometimes became full of developmental envy and jealous of their superior officers. They evidently at times even refused to cooperate with their orders so that their commanders would lose their battles and their jobs. Many soldiers lost their lives as a result.

And at the holiday gathering, maybe we sit down to eat the turkey dinner. As one of the adults passes the mashed potatoes, they make a snarky political comment, knowing from many years of experience that this will annoy at least half of their relatives in the room. And, although now fully cooked, the turkey flies again, across the table.

Just last week my yoga instructor addressed this same issue we are contemplating. We students were all aligned on our mats when he sat down and asked, "Why do we do these exercises?" Then he answered his own question: "Instead of wishing that this stressful world was different, we can teach our bodies to relax and our minds to be present."

I like that idea; it reminds me of how famous author Victor Frankl wrote that the "last of the human freedoms" is to "choose one's attitude in any given set of circumstance" (Goodreads, 2002). We are all free to stay in polarization with other people or to not.

Most of the news we are fed today is tough to hear. We have received over the years increasingly dire warnings about the climate crisis, with daily updates on record-breaking temperatures, fires, and flooding. With the Russian-Ukraine war and increased tensions with China over Taiwan, talk of nuclear war has started up again. Other wars, equally lethal and dangerous, have gone on for years in Asia, Africa, and the Middle East, although they have received less attention in our Western news. Racial, religious, and nationalistic differences regularly become reasons for unnecessary conflict and violence. Then, on October 7, 2023, Hamas attacked Israel, Israel retaliated, and we have received regular bulletins about this new war ever since.

To make matters worse, the challenge of polarization seems to have taken hold of humanity; how can we make progress in addressing all these global challenges when we seemingly cannot cooperate?

What is *polarization*? The verb *polarize* has its origins in the physics of optics, from the early 17th century, derived from the Modern Latin *Polaris,* which refers to the two "polar" ends of the planet. More recently in the mid-20th century, the word *polarization* took on an additional meaning: "to accentuate a division in a group or system" (Online Etymology Dictionary, n.d.).

Are we accentuating and exaggerating our human differences and minimizing our similarities? And if so, why? In *The Identity Trap: A Story of Ideas and Power in Our Time* (2023), Yascha Mounk argued that in the United States today, the identities that each of us have increasingly become *the* focus of our cultural, social, and political lives. Mounk calls this centering of personal identities the "identity synthesis":

> The identity synthesis is concerned with many different kinds of groups, including (but limited to) those based on race, gender, religion, sexual orientation, and disability. It is the product of a rich set of intellectual influence, including postmodernism, postcolonialism, and critical race theory. ... Many advocates of the identity synthesis are driven by a noble ambition: to remedy the serious injustices that continue to characterize every country in the worlds, including the United States. (pp. 9–10)

After clarifying that he strongly supports efforts for equity, diversity, and inclusion in our institutions, communities, and social systems, Mounk goes on to explain:

> Far right ideologies are so dangerous because they discourage people from widening their circle of sympathy. ... Placing specific ethic or cultural identities on a pedestal, they encourage their followers to value their group over the rights of outsiders or the claims of universal human solidarity. ... [The] identity synthesis ... [also] makes it harder for people to broaden their allegiances beyond a particular identity in a way that can sustain stability, solidarity, and social justice. ... [It is] likely to create a society composed of warring tribes rather than cooperating compatriots, with each group engaged in a zero-sum competition with every other group.. The identity synthesis is a political trap, making it harder to sustain diverse societies whose citizens trust and respect each other. It is also a personal trap, one that makes misleading promises about how to gain the sense of belonging and social recognition that most humans naturally seek. (pp. 13–14)

Another important current influencer, Indian author Pankaj Mishra (2017), has addressed the topic of anger and polarization from a broader, more global perspective. In *Age of Anger: A History of the Present* (2017), Mishra attributes our current social and political divides and polarization to a long-term global reaction to industrial capitalism and modern liberalism in the Western world. He identifies a near-universal reaction of "resentiment" that exists across the world, in which many of us blame other populations of people for our own suffering.

Mishra explains that people increasing feel justified to react to the consequences of capitalism and liberal democracy with anger and violence. According to Mishra, the economics and politics of the West have undermined traditional families and communities across the world. Extreme political movements in both the East and West are then increasingly motivated by these large-scale dynamics.

From Mishra's perspective, our current brand of polarized politics no longer serves those who are oppressive and disadvantaged: "Politics now is really only about self-interest, which means it has violence built into it because your self-interest is going to collide with the self-interest of the rest of the world."

He reminds us that the rest of world is both susceptible and skeptical of the solutions offered by wealthier nations to poorer nations: "There is this idea of history as something you make, as a meaningful narrative with a beginning and an end, the end being a utopia of happiness that we'll reach through socialism or free trade or democracy, and then it will all be wonderful" (AZquotes, n.d.).

A commonality in the perspectives of both Mounk and Mishra is the idea that the extreme individualism and self-interest associated with our Western culture, economies, and politics can ultimately result in harm to individuals and societies. Although we will not further explore the dangers of extreme *collectivism* in this chapter, it is worth noting that, as we have examined in our text, *both* extreme collectivism and extreme individualism are potentially dangerous, and both can undermine our attempts to cooperate in addressing our climate crisis.

We can conclude for now that polarization involves not just exaggerated self-interest or the accentuation of differences between individuals, groups, families, communities, and nations, but also the lack of civility about these differences. Polarization involves how people think about, speak, and act toward each other, and all these symptoms are interrelated. We have explored in our text how the individual and collective ego, when unexamined, can exaggerate my small self-interest, identification with my beliefs, and dehumanization of other people who look, act, think, or live differently from how we do. Simply said, when

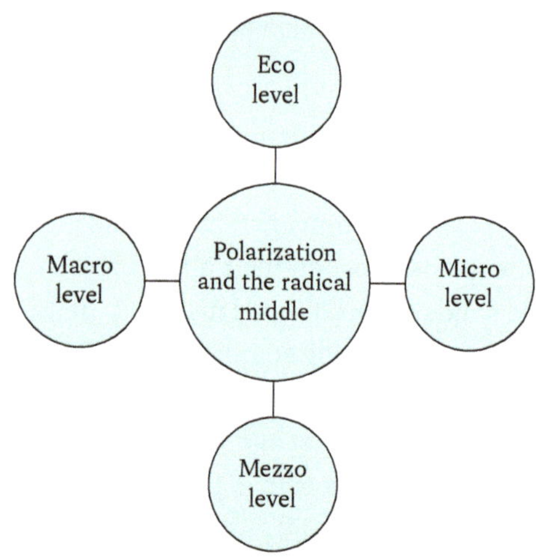

FIGURE 7.1 A Model of Four Levels of Practice With Polarization and the Radical Middle

I identify with my own beliefs and dehumanize another group for any reason, my mind can justify violence toward others.

Exploration of the Micro, Mezzo, Macro, and Eco Levels of Polarization

In this chapter, we will examine the etiologies of and possible solutions for this polarization, on all four levels of analyses: the micro, mezzo, macro, and eco. This analysis will help us better understand the complex associations between many factors, including extreme individualism, identity, ego, polarization, politics, the economy, and the climate crisis in our world today. In Figure 7.1, we build on Figure 1.2 in Chapter 1 to illustrate the interaction between the micro, mezzo, macro, and eco levels and polarization.

In addition, from an inclusive practice perspective, there is always a middle ground, or a *radical middle* position, from which we can seek the truths that usually exist on both sides of a disagreement. We will seek to find a radical middle perspective that can help heal and transform polarization. We will begin to build a radical middle position on the politics of climate crisis that can potentially help us make the progress we now need to meet the challenge we now face today.

Finding the Radical Middle Through Micro-Level Practice With Polarization

What does polarization look like at the micro level? There is evidence that relationships many of us have in our families and communities are affected by how we currently react to political and other ideological differences. For example, about 80% of Americans now have "just a few" or no friends at all who vote for politicians across the aisle from the people they support, and many of us unfortunately have relationships in which we experience "dualling realities" with our family members, neighbors, and friends (Smith, 2020). According to cultural psychiatrist Charles Johnson (2021), "Extreme polarization is setting neighbor against neighbor and often directly putting us at risk." Other research, however, suggests that many of us are still willing to care about family members who vote differently than us and to tolerate a diversity of ideas in our relationships (Abrams, 2022).

Researchers have found evidence that the beliefs that divide us may be associated with personality characteristics that seem to have at least some genetic etiologies. Hetherington and Weiler (2018) found that individuals in our culture tend to adhere to one of three types of world views. The "fixed" (more conservative) view is associated with the values of respect, obedience, good manners, and good behavior and tends to see more danger in the world. The "fluid" (more progressive) world view especially values independence, self-reliance, curiosity, and being considerate and tends to view the world as a safe place to explore. The "mixed" world view has aspects of both fixed and fluid views. These personality types, which tend to persist across time, seem to be strongly associated with our political views.

Helping professionals can assist our clients in understanding that the ideologies people have are associated with such lifelong personality traits and that many of us also often react due to our fear and anxiety when we identify with particular political or ideological stances. From a radical middle perspective, we can also consider the probability that these personality differences evolved for a purpose and that every family and community and country needs people of all three personality belief types to best survive and prosper. We need people in every community, for example, who are willing and able to be brave warriors, as well as others who are willing and able to be compassionate spiritual teachers, others who are willing and able to be wise leaders, and so on. In other words, such diversity contributes to community well-being when the community honors such diversity.

A professor who was an important mentor of mine used to say, "People do the best they can under the circumstances as they see them." We can also remind our clients of the wisdom of this perspective as we strive to teach them how to work and live effectively and cooperatively with people who think differently from how they do. In doing so, we reaffirm the humanity of other people who see things differently and refuse to only see them as evil or bad. Instead, they are doing the best they can.

Another intervention we can use with polarization, when we are seeing a client individually or with family members, is to work with their consciousness. As discussed in Section 1, we humans have an ego that often is dissatisfied with the way things are in the world and makes protests, such as "this should not be happening" or "this is not fair." The helping professional can help the client see how these protests are typically ego-based and tend to lead to unnecessary suffering. As the client becomes conscious of how their ego is operating, the ego no longer has the power to control their behavior, and the client accepts the world the way it is—full of people who may or may not disagree with us. We could say that polarization decreases as consciousness increases.

I have found that when there is polarization in small groups, such as in families, work locations, or neighborhoods, developmental envy is often associated with the hostility between people. What is *developmental envy*? I mentioned the idea in the introduction to this chapter. It is a common reaction, for example, that I might have to someone else in my life who seems more advanced in some area of growth than me. In other words, I might be jealous of you because you seem to know more than me, you seem more popular, you appear more confident or beautiful, and so on. The reason such envy can lead to polarization is because we are often embarrassed that we have such feelings and therefore try to hide them from others and ourselves. And when we deny our feelings, we cannot heal and transform them, and therefore, we try to find other reasons to justify our often intense discomfort and aggressive reactions. So, instead of recognizing that I am envious, I might, unfortunately, project all kinds of additional negative attributes onto my loved one, colleague, or neighbor. Professional helpers can work with our clients to assist them in becoming more conscious of their developmental envy so that they do not have to create unnecessary suffering for themselves and others.

In general, we could explore briefly here an important aspect of practice with polarization. The question is, is it best to ask our clients, students, or patients to first focus on their internal work of increasing consciousness of their individual and collective ego functions, or does it make more sense to begin with a focus on changing behaviors? The short answer is "It depends."

It depends upon the context of the situation and the motivation and abilities and life-context of the people we are working with. Perhaps surprisingly, it is often easier to act my way into new ways of thinking than to think my way into new ways of acting. This is because many of us are already so much in our heads (i.e., mentally focused) in our culture. We can leap past our habitual patterns of thinking sometimes by first changing our behaviors and, for example, start treating other people differently. Behavioral changes can then lead to more openness in looking at our deep internal dynamics. However, change can also go the other way, and consciousness work can certainly often lead to changes in behavior. Some people prefer doing inner and outer work at the same time. We recommend experimenting with your clients, students, and patients to find what works best for them.

Dialogue can also be a very effective practice approach in working with polarization at the micro level. As part of the dialogues for bridging the religious divide, mentioned at the beginning of the chapter, for example, I often watched people learn how to be in a relationship with each other as they shared their diverse religious, political, and other ideological beliefs. We can create frequent and regular opportunities for dialogue within our communities, where people can meet together to talk and build relationships.

Perhaps the most important dialogue ground rule we listed in Chapter 6 is to listen for understanding. Lacy, a Ute Medicine Man and one of my favorite colleagues with whom I worked at the Indian Walk In Center, often said in group, "Creator gave us one mouth and two ears for a reason. Make sure you listen two times more than you speak." I have seen many of my students, clients, and colleagues become more compassionate and accepting of others after they learned to listen for understanding in difficult dialogues.

Finding the Radical Middle Through Mezzo-Level Practice With Polarization

At the mezzo level of practice, we may work with communities, institutions, other work locations, or organizations. A polarized community is one in which the differences between people are highlighted and exaggerated—where there is increased lack of trust, fear, and hostility between groups and where people may work and live in increasingly segregated groups.

There are many forms of polarization, as many forms as there are ways for humans to be different. In the United States, we often talk about *political* polarization, which Jesse Shapiro (Brown University, 2020) calls "affective polarization." Shapiro and colleagues found in their study that since the late 1970s, affective polarization in the United States has increased more quickly than in Canada, the United Kingdom, Australia, New Zealand, Sweden, Germany, Switzerland, and Norway. Although they did *not* find that the Internet was particularly associated with the polarization increase in the United States, the researchers *did* find that increases in 24-hour partisan news and decreases in public broadcasting were especially prevalent in the United States (compared with the other countries examined) during the years studied. These findings suggest that increased access to *nonbiased* news may reduce our political polarization and that we may also need to publicly examine and expose the domestic and foreign interests that often support partisan news.

There are multiple approaches that we can use to intervene with mezzo-level polarization. One approach is to focus upon leadership skills. Over the years while working with and in various organizations and communities, I have found that top leaders set a tone that can be felt throughout the group for which they work. Professional helpers therefore often need to work with key leaders to foster change. The Harvard Business School (Landry, 2018) has provided descriptions of effective leaders over the years, and Table 7.1 summarizes content from online material it has developed.

TABLE 7.1 Characteristics of Effective Leaders

Characteristics	Brief Descriptions
1. Ability to influence	Able to listen and establish trust
2. Transparency	Balance transparency and privacy
3. Encourage innovation	Support experimentation
4. Integrity and accountability	Balance power and accountability
5. Decisive actions	Know when to act, evaluate, pivot
6. Resilience	Responsibility, optimism, flexibility

Source: Landry, 2018.

If I were on a hiring committee for a new community or organizational leader, I would first look for self-awareness and self-acceptance in candidates, because I see self-awareness and self-acceptance as foundational for all six characteristics listed in Table 7.1. The same goes for voting for elected officials. And, as we have explored throughout the text, these two traits, awareness and acceptance, are equivalent to what we are calling *consciousness*. In other words, more conscious leaders make more effective leaders who can help us reduce polarization in our communities.

Dialogue can also be effective in helping to transform the culture of a community or organization, as we described in Chapter 6. Communities and organizations sometimes will sometimes reach out for help in changing their culture, and professional helpers can offer dialogue facilitation as an approach to supporting such transformation. The setup makes a difference in the dialogue's effectiveness.

In working with culture (climate) change, I have learned to meet first with leaders to find out what their concerns and goals are. It usually helps to have the key leaders set the tone for a dialogue by welcoming participants at the time of a dialogue and explaining to them its context. When dialogues are mandatory, there is usually better attendance but also some resistance from some participants; when dialogues are voluntary, there may be fewer people, but often they are more motivated. Sadly, the people who are most willing to participate are often the folx who need the dialogue the least, but we can only work in our dialogues with those people who attend them.

When possible, I have enjoyed co-creating a dialogue facilitation team with whom to work. For example, with the help of a grant from my university, I created a program years ago for the Center for Teaching and Learning Excellence that I called "Transforming Classrooms Into Inclusive Communities." We co-created a facilitation team and were extremely busy for several years during the grant period, working with departments and colleges on campus that asked us for assistance. We would always meet first with leaders in each unit, listen to their stories, and describe what a dialogue is and how it might help. We often found that administrators did not understand the difference between a dialogue and a debate, so those variances needed to be explained.

We would always ask for a room, if possible, that had open space for chairs to be arranged in circles and avoided rooms with rows of seats and tables fastened to the floor. Often there were questions or topics that the leadership team wanted us to cover. During the actual dialogue, we would use a combination of large- and small-group conversations, depending on the group's estimated size. We found that when the number of participants was 25 to 30 or fewer, we could stay in a large group throughout the dialogue. When the number was higher than 30, people usually found that they benefited from meeting for part of the time in smaller breakout groups so that more people could have varied opportunities to speak and listen.

Most participants in these programs reported that they had found the dialogues useful in creating relationships across the differences in their unit that previously divided them. Many of the units went on to ask for more dialogues, during which they began to find more and more common ground. We usually found that multiple dialogues typically helped foster deeper conversations as mutual trust developed.

Finding the Radical Middle Through Macro-Level Practice With Polarization

Although most people (about 71%) are concerned about the direction of our country, there is little consensus yet about the nature of our challenges and the directions we need to take together in response (Todd et al., 2023). However, there are important conversations going on now in the United States that may help us find the radical middle in the polarization that seems to be constricting us.

David Brooks (2023), a frequent contributor to *The New York Times* editorial page, has critiqued the tendency, especially seen on "elite" U.S. college campuses, to emphasize only the differences between people, see all relationships as power struggles, and minimize the ability of people to understand each other and transcend the differences that divide us.

Although he has been criticized by other journalists (e.g., Lehmann, 2023) for ignoring the role that the "capitalist political economy" plays in creating the increased polarization and self-interest so prevalent today, other writers from across the political spectrum have recognized Brooks's insights into the etiology of polarization. For example, Damon Linker (2019), writing for *The Week*, applauded Brooks's notion that "anti-pluralism" is in fact the "challenge of our time." He writes:

> Since Donald Trump won the GOP nomination and then the presidency in 2016, the political tendency he represents has been called many things. At first many labeled it populism. Then, with the blessing of the president and his advisers, commentators fastened onto nationalism. Meanwhile, his political opponents have insisted all along on describing it as an expression of white supremacy and perhaps even the leading edge of fascism. ... There's some truth in all of these terms, but I think *New York Times* columnist David Brooks does better when he describes the political mood of the moment as "anti-pluralism," which he defines as a reaction against "the diversity, fluidity, and interdependent nature of modern life." As Brooks goes on to explain, "Anti-pluralists yearn for a return to clear borders, settled truths, and stable identities."

We dialogue facilitators want to give a voice to all members of the community, including those who feel threatened by changes that many of us may believe is progress. I have found that when we listen to people talk about their concerns about social change, their fears tend to lesson.

Professional helpers can take a radical middle position when it comes to any conversation about "what is wrong with our country." The radical middle position allows us to hear voices from across the political spectrum so that we can stay in relationship with our neighbors. A strong defense against threats to democracy, uncivility, and polarization may well be a society of inclusive communities that resist polarization by welcoming all voices and encouraging ongoing community dialogue. As Mounk (2023) points out, "demagogues thrive when societies are deeply polarized and decision makers are out of touch with the views of average citizens" (p. 18).

Relationship building and consciousness building are skill sets that political leaders also may need to practice. A study conducted by Kim Davenport and her students at Vanderbilt University (McGuire, 2021) found that consensus exists between politicians regarding the causes and remedies for political polarization. The respondents agreed that *collaboration* is the most important competency for building consensus, followed by integrity of character, empathy, courage, and willingness to balance principles and pragmatism. Barriers to effective collaboration began with *self-interest*, followed by party partisanship, political opportunism and extremism, reelection pressure, and the perception that compromise is a sign of weakness. Their advice for overcoming these barriers started with *genuine relationships*, followed by balancing principles and pragmatism, making systemic changes in such items as finance reform and term limits, and increasing public support for collaborative politicians.

Professional helpers can teach our students, clients, and communities to strive to stay in relationship with our families, colleagues, and neighbors. *To be in relationship* with another person means to continue recognizing, communicating with, and caring for that person. We do not have to pretend to like everything about them, but we can look for their strengths and likeable qualities as we interact.

Collective Ego

In our discussions about ego, we saw how it can take both an individual or a collective form. *Collective ego* might be defined as the identification of an entire group of people with a belief or set of beliefs, associated with the view that people who share my particular political system, national identity, religious belief, and/or cultural or racial identity are superior (or inferior) to everyone else.

When working at the macro level, we understand that any group of people can have a collective ego, from a small family group to a local neighborhood or community to an entire state or nation. Remember that there is nothing wrong, of course, with loving my political system, country, religion, culture, or race. What we are concerned about is what happens when there is an identification with my beliefs, which, unfortunately, tends to result in my dehumanization of people who hold different beliefs. When we dehumanize the Other, we make it possible to feel resentment and act violently toward outgroup people. Thus, collective ego can result in polarization, because I feel the need to protect my group identities from all perceived threats from other identity groups.

For example, I can love the country in which I live and be very proud of many things about my community while at the same time refusing to dehumanize people who live in a different location or who hold different beliefs or characteristics from those of the majority group to which I belong. In addition, when I love my country, I can still see the imperfections that exist and strive to address them. Love is not blind, but rather strongly associated with consciousness; my reverent awareness enables me see the world the way it actually is so that I can love *what is*, rather than what I might think *should be*.

Thus, the "medicine" for collective ego is the same as for individual ego, and that is our consciousness. When I am aware of my egoic nature and tendency to identify with what the collective ego identifies with,

I become free to see myself and the world accurately. I can find the radical middle and see and love people for what we actually are, as humans with a mixture of strengths and limitations.

Please note that we use the word *identity* in two ways throughout the text because the word is used in multiple ways in the English language. The first two primary meanings of the word *identity* listed by MerriamWebster (2023) include (1) "the distinguishing character or personality of an individual" and (2) "the condition of being the same with something described or asserted." These are the two ways we use the word in our text as well. When, for example, we talk about how Mounk (2023) primarily uses the term *identity* in his book *The Identity Trap: A Story of Ideas and Power in Our Time*, we are using the first meaning. When we discuss how Ekhart Tolle uses the term *identity* or *identification* in his book *A New Earth*, we are using the second meaning. So, it is possible to say in the same sentence that a person or group may identify (here, we're using the second meaning) with any of the identities (the first meaning) that they may happen to have. However, I have strived to avoid using both meanings of the word in the same sentence to reduce confusion. Perhaps we English speakers could agree to come up with a different word for one of the two meanings, to reduce future confusion; for example, we could use the word *characteristic* to replace *identity* for the first meaning. The two meanings actually frequently overlap in our real-life experiences and usages, which can add to the confusion.

Role of Capitalism

When working at the macro level, we can also examine and respond to the ways in which our current state of polarization is associated with our economic and social humansystems. I find that Naomi Klein's (2023) analysis of the relationship between our capitalistic economic system and the great challenges of our time to be very compelling. She writes about:

> capital's ravaging of our bodies, our democratic structures, and the living systems that support our collective existence. ... We are now reaping the rotted harvest of decades of deliberately sown mistrust—mistrust of the very idea that we are members of communities and societies, mistrust of any expectation that government can and should do anything positive for us. (pp. 230–232)

Klein adds:

> It took decades to desocialize people to accept neoliberalism's cruelties. Anti-Black racism and anti-migrant hysteria played powerfully enabling roles with a steady stream of politicians and media titans equating social programs designed to help everyone in need with Black "welfare queens," "super-predators," and "illegals." ... The legacy of generations of messages that pitted members of society against one another does not disappear overnight because there is a pandemic. (p. 233)

(And, of course, we could add that they have not disappeared overnight because of our ever-increasing climate crisis either.)

Klein goes on to explain that the conspiracy culture that so often now fuels polarization tends to ignore the U.S. "hyperindividualism" that blames individuals (rather than systems) for their economic struggles and, in fact, reinforces that individualism by often singling out powerful figures for the country's problems. The COVID-19 pandemic revealed to us that there are macro-level issues that require us to see each other and cooperate (rather than blame each other) to get things done. Our climate crisis, of course, does the same thing. In other words, there is actually a *real* conspiracy, one that exists within capitalism. According to this concept, an elite class of the wealthy and powerful exists who encourage us to be polarized so that we become more susceptible to imaginary conspiracies about which we can fight with each other rather than focus together on the already severe and still-growing lack of income, wealth, and power equity that capitalism promotes in the United States and most of the world (Klein, 2023).

One of the most effective ways for professional helpers to resist this conspiracy of division is to organize public dialogues that bring people together so that we can begin to bridge the differences that currently separate us. In such conversation, we can, yes, acknowledge our differences; however, we also can see our common interests and engage in faith-based activism that can support the highest good of our local and global communities. Some dialogue groups might eventually decide to take some collective actions against corporations that they want to influence, such as refusing to buy products or services from them. Although such local work may seem tiny compared to the massive economic and political power of some corporations, if all or even many local communities held such dialogues and engaged in follow-up activism, much of the power of those corporations that benefit from and encourage polarization could be diminished.

We can also help educate our community members to be aware of when leaders seek to increase our fear of and judgment and anger toward any other group of people. As Klein and many others point out, we have many examples from history of how this approach has often led to organized violence toward others, through oppression, enslavement, war, and even genocide. As we will discuss further in Chapter 10, we do not have to look very far to see examples of such violence in our current world happening right now.

Finding the Radical Middle Through Eco-Level Practice With Polarization

Finally, we will use the phrase *eco-level* (see Figure 7.1) to describe our work that involves people's interactions with other living things and the ecosystems that support all life. Examples of eco-level work include:

1. Interventions with farmers dealing with the consequences of severe drought
2. Meetings with small-town residents whose community was destroyed by wildfires
3. Work with Indian tribal members who are struggling to have their water rights honored by the state
4. Community organization with urban homeowners concerned about the impact of polluted air in winter-valley inversions on the well-being of their children
5. Practice with ranchers who live near a national park that protects the growing wolf population

In each of the examples given here, there may be eco-level polarization that complicates the work that professional helpers want to do with these populations at risk. For example, farmers who have lost their crops to drought or are facing possible severe losses may, in many cases, feel unsupported and even undermined by government agencies as well as by other populations. Victims of wildfires may also feel similarly isolated, misunderstood, and ignored. Indian tribal members may feel that they are invisible to the rest of the country and may believe that the majority of people in the country do not care about and even actively hate them. The homeowners who deal with inversions and severe air pollution may resent the owners of oil refineries and other major polluters in their valley who make the air quality worse. Finally, the ranchers who lose some of their cattle to the wolves may believe that the people seeking the wolves' protection are elite urban environmentalists who do not care to understand how ranchers feel.

Researchers have identified some paths toward depolarization that have proven successful in such situations. Bradbery and Johnston (2023) suggest some social change strategies that have been proven successful, including efforts to mobilize people to cooperate together to create positive social change through dialogue techniques. They suggest that facilitators can help both sides use active listening skills, think about the issues and each other in nonbinary terms, and locate shared common experiences and common values.

I have found that nature can often draw people together because most of us support such things as natural habits for wildlife, national parks, city parks, clean air and water, and beautiful natural views. For example, a recent poll shows that about 90% of those surveyed say more needs to be done in our national parks to protect fish, birds, animals, and their habitats (Repanshek, 2023). Facilitators and educators can help people find the radical middle by seeking the shared values and experiences that can move us toward greater trust, empathy, and cooperation.

There is an art to the way in which we create our messaging for the public. As Anand Giridharadas (2022) has offered, for example, it is often more effective to "paint the beautiful tomorrow" rather than to complain about what we are against. Said differently, he adds, "What you fight, you feed." We are more effective in dealing with our climate crisis when we talk more about what we stand for rather than solely what we stand against. Instead of beginning our public conversations, classes, and therapy sessions with what we are angry about and whom we are angry at, it may be more effective to first discuss shared values, identify the common challenge we face, and move toward finding possible solutions. By emphasizing our own anger, we may increase polarization, but by highlighting our commonality, we may help move us toward that radical middle ground where everyone can "win" a little.

The well-being of society is more important than political victories. And the well-being of other living things and of the ecosystems that support life is essential to the well-being of society. We helping professionals can remind people of these simple truths.

We can help people see that transcending polarization is about growing up. Adults see beyond their own immediate self-interest; understand that there is always a greater, common good; and take responsibility for cooperating with other people for that common good.

QUESTIONS FOR REFLECTION

1. How have you experienced polarization in your own life? Can you identify examples of polarization you have experienced at the micro, mezzo, macro, and eco levels?
2. After reading Chapter 7, think about which areas of micro-, mezzo-, macro-, and eco-level practice you are most interested in and which ones you feel more competent in pursuing. Often the work we enjoy is also the work we are good at. Is that true for you? Please explain.
3. Can you think of a climate crisis issue that our society is currently polarized about, that you would like to help us all move toward a new radical middle? Please explain. How would you help us with this shift?

4. We have suggested in this chapter that it can be more effective to identify the futures that we hope for rather than what we are angrily opposed to now. Please identify a future dream you stand for and explain why.

REFERENCES

Abrams, S. J. (2022). *Polarization in American family life is overblown*. Survey Center on American Life, February 23, 2022. https://www.americansurveycenter.org/polarization-in-american-family-life-is-overblown/

AZquotes. (n.d.). *Pankaj Mishra quotes*. https://www.azquotes.com/author/52792-Pankaj_Mishra

Bradbery, A., & Johnston, J. (2023). A path toward depolarization. *Stanford Social Innovation Review, 21*(4), 61–62. https://doi.org/10.48558/F357-M842

Brooks, D. (2023). Failing at inclusion. *The New York Times*, November 16, 2023.

Brown University. (2020). *U.S. is polarizing faster than other democracies, study finds*. January 21, 2020. https://www.brown.edu/news/2020-01-21/polarization

Giridharadas, A. (2022). *The persuaders: At the front lines of the fight for hearts, minds, and democracy*. Knopf.

Goodreads. (2002). https://www.goodreads.com/quotes/1287210-the-last-of-the-human-freedoms-to-choose-one-s-attitude

Hetherington, M. J., & Weiler, J. D. (2018). *Prius or pickup? How the answers to four simple questions explain America's great divide*. Mariner Books.

Johnson, C. (2021). Today's extreme social and political polarization. *Psychology Today*, July 15, 2021. https://www.psychologytoday.com/us/blog/cultural-psychiatry/202107/today-s-extreme-social-and-political-polarization

Klein, N. (2023). *Doppelganger: A trip into the mirror world*. Farrar, Straus and Giroux.

Landry, L. (2018). *Six characteristics of an effective leader*. Harvard Business School. https://online.hbs.edu/blog/post/characteristics-of-an-effective-leader

Lehmann, C. (2023). The patronizing moralism of David Brooks. *The Nation*, August 16, 2023. https://www.thenation.com/article/society/david-brooks-moralism/

Linker, D. (2019). *The rise of anti pluralism is the challenge of our time*. The Week, August 8, 2019. https://theweek.com/articles/857582/rise-antipluralism-challenge-time

McGuire, M. (2021). *Political leaders agree on core attributes needed to reach across the aisle*. Vanderbilt Project on Unity and American Democracy, December 22, 2021. https://www.vanderbilt.edu/unity/2021/12/22/political-leaders-agree-on-core-attributes- needed-to-reach-across-the-aisle/

Merriam-Webster. (2023). *Identity*. https://www.merriam-webster.com/dictionary/identity

Mishra, P. (2017). *Age of anger: A history of the present*. Farrar, Straus and Giroux.

Mounk, Y. (2023). *The identity trap: A story of ideas and power in our time*. Penguin.

Online Etymology Dictionary. (n.d.). *Polarization*. https://www.etymonline.com/word/polarization

Radical Middle Newsletter. (n.d.). Radical middle: The book: Q & A with author Mark Satin. http://www.radicalmiddle.com/book_Q&A.htm

Repanshek, K. (2023). *National poll shows Americans' strong support for wildlife*. National Parks Traveler, November 17, 2023. https://www.nationalparkstraveler.org/2023/11/national-poll-shows-americans-strong-support-wildlife

Smith, T. (2020). 'Dude, I'm done': When politics tears families and friendships apart. National Public Radio, *All Things Considered*, October 27, 2020. https://www.npr.org/2020/10/27/928209548/dude-i-m-done-when-politics-tears-families-and-friendships-apart

Todd, C., Murray, M., Bowman, B., & Marquez, A. (2023). Poll finds 71% of Americans believe country is on wrong track. NBC News, January 30, 2023. https://www.nbcnews.com/meet-the-press/first-read/poll-finds-71-americans-believe-country-wrong-track-rcna68138

Credit

IMG 8.1

Climate Justice

Practice in Diversity, Equity, and Inclusion

When will our consciences grow so tender that we will act to prevent human misery rather than avenge it?

— ELEANOR ROOSEVELT (N.D.)

You cannot know your brother when you attack him. Attack is always made upon a stranger. You are making him a stranger by misperceiving him, and so you cannot know him. It is because you have made him a stranger that you are afraid of him. … Because your attack thoughts will be projected, you will feel attack. Projection and attack are inevitably related, because projection is always a means of justifying attack. … Safety is the complete relinquishment of attack. … The strong do not attack because they have no need to do so. Before the idea of attack can enter your mind, you must have perceived yourself as weak.

—FRANCES VAUGHAN AND ROGER WALSH (1988, P. 152)

It is the first day of December 2023. As I type these words, a crowd of delegates, government representatives, and world leaders are attending the 28th annual United Nations Climate Change Conference (COP28) in Dubai, United Arab Emirates. There have been calls again this year for wealthier countries to take more responsibility for our disproportionate role in creating the climate crisis and provide sufficient funding and other assistance for poorer countries as they struggle to deal with climate extremes and finance transitions to new energy sources (Gelles & Walt, 2023). As global inequality in power and wealth continues to increase, the disproportionate degree to which the wealthiest generate greenhouse gas (GHG) emissions also increases: The richest 1% now produces twice the carbon emissions of the poorest 50% (Harrabin, 2021).

One of the biggest challenges in humanity's efforts to deal effectively with the climate crisis exists in our own xenophobia. Xenophobia, or the fear of people who appear to think, act, or look differently from us, is a driving force behind our lack of cooperation with people who may look or act differently than us. It is also associated with the lack of diversity, equity, and inclusion (DEI) in our local and global humansystems today, and people who are already minoritized are often also more at risk of experiencing negative impacts as our climate crisis worsens.

When we seek DEI, we seek *social justice*. As we move toward greater climate justice, we develop a more *diverse* group of leaders who represent all the voices in our local and global community, create more *equity* in how much every population contributes to positive solutions and benefits from those solutions, and co-create more *inclusive* local and global communities that welcome all populations to share in Earth's resources. The promotion of climate justice helps transform our humansystems, creating the cooperation necessary to effectively deal with our climate crisis.

Xenophobia is not the only factor that drives climate injustice, of course. As we explored in Section 1, the human ego, rooted in the left hemisphere, is associated with xenophobia as well as with our *grandiosity*, *disconnection*, and *desire for power, wealth, and fame*. Figure 8.1 illustrates how these *egoic* processes are all drivers of climate injustice in equity, diversity, and inclusion. In Section 1, we explored how our work with human consciousness can help heal and transform grandiosity and the desire for power, wealth, and fame. In Chapter 8, we will focus primarily on the healing and transformation of xenophobia.

How are these egoic processes associated with climate injustice? In short, our xenophobia is associated with our inability to see the Other as fully human and with our tendency to ignore their needs and attack them. Our grandiosity, or tendency to feel superior to others, is associated with our sense of disconnection, entitlement, and insensitivity to inequality. And our desire for power, wealth, and fame easily can lead to a tendency to take all we can from both other people and from nature, and then to fight to hoard and retain what we have gained.

Climate justice addresses the fact that our climate crisis affects different populations differently and that our existing inequality is often made worse by the disproportionate impact on vulnerable populations of such climate extremes as severe heat or cold, drought, fire, and flooding. Since the climate crisis negatively impacts or will probably eventually impact almost all people on the planet, we advocate for the protection of all populations, regardless of their wealth, power, or status. However, we are especially sensitive to those populations who are most at risk in each unique situation.

Table 8.1 summarizes examples of some of the major populations at risk who are especially vulnerable to climate crisis events, according to Yale Climate Connections (Simmons, 2020). The population categories used in the table can sometimes overlap, and additional details on populations at risk are presented later in the chapter.

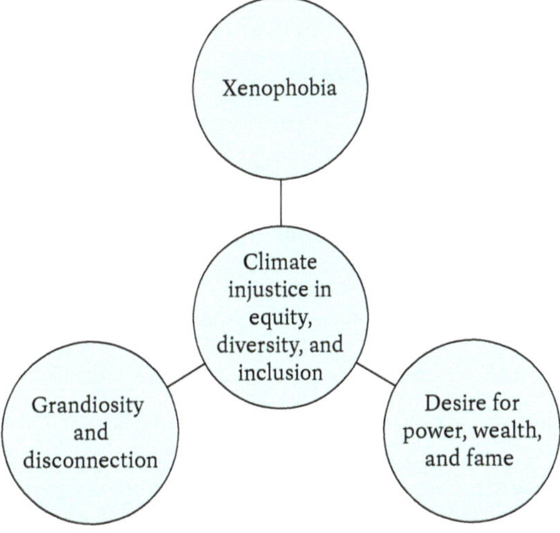

FIGURE 8.1 The Egoic Processes of Xenophobia, Grandiosity, and Desire for Wealth, Power, and Fame Are All Drivers of Climate Injustice in DEI

TABLE 8.1 Examples of Populations at Risk That Are Especially Vulnerable in Our Climate Crisis

Some Populations at Risk	Climate Crisis Events	Examples of Possible Climate Vulnerabilities
Communities of color	Air pollution Extreme heat	Live and work in vulnerable environments Have few resources to bolster resiliency Are unable to afford safer living sites Are unable to repair damages Face rising food prices and other living expenses
Aging populations	Severe heat Fire flooding	Are at risk for physical complications Are unable to evacuate
People with disabilities and chronic illness	Heat waves Fires and flooding	Are at risk for physical complications Are unable to evacuate
Communities of poverty	Heat waves Fires and flooding	Live and work in vulnerable environments Have few resources to bolster resiliency Are unable to afford safer living sites Are unable to repair damages Face rising food prices and other living expenses
Immigrants	Heat waves Fires and flooding	Are unable to communicate Have few resources to bolster resiliency Work in agriculture and other outdoor work
People with language barriers	Heat waves Fires and flooding	Are unable to communicate Have few resources to bolster resiliency Face social isolation
Indigenous populations	Sea-level rise Drought flooding	Live and work in vulnerable environments Have few resources to bolster resiliency Are unable to afford safer living sites Are unable to repair damages Face rising food prices and other living expenses
Youth	Climate crisis	Deal with physical and mental health impacts Deal with long-term trauma and other vulnerability

Xenophobia has also been closely linked to violent conflict and war throughout human history, and war is itself often very destructive to all living things and ecosystems. As most of us probably remember from our middle and high school classes, human history is often recorded as ongoing stories of conquests of one group of people over another. Arguably driven at least in part by xenophobia, these conquests are typically contests for power, wealth, and fame. The invaders usually regard their victims as inferior and seek to dehumanize them (*National Geographic*, 2023).

The history of human interaction with nature is unfortunately also full of similar stories of conquest, except that the conquests have been over other livings and the ecosystems that keep us all alive. Perhaps at least in part because of our fear of nature, most human beings came to view the natural world as inferior to the human world and have strived to have mastery and control over it (Frankopan, 2023). In our text, we will use the term *biophobia* to describe *all* kinds of fear about nature. We will explore these fears in this chapter and later in Section 3 when we discuss our relationship with ecosystems.

Our progress in addressing the climate crisis—indeed, in addressing any global challenge—requires us to address the xenophobia that humanity still harbors today. This is primarily because human cooperation is required for us to deal effectively with the complex local and global challenges in the climate crisis. Our long-standing tendency to mistrust, fear, dehumanize, and attack people who we see as different therefore now threatens the survival and well-being of current and future generations.

These xenophobic qualities of humankind can be sobering and frightening to contemplate, but healing and transformation are always possible. We can embody a faith-based activism because there is also evidence that altruism is also part of most humans. In Chapter 4, for example, we examined how the archetypes of Eros and Thanatos both can be expressed for the highest good. According to Harvard Medical School (Miller, 2016), "altruism or conscience [is] a naturally emerging property of human societies. We humans do seem to have the capacity to play fair, even if we sometimes need the fear of punishment to help us do so" (p. 1). We can foster the further development of our altruism and reduce our tendency toward xenophobia.

In the remainder of this chapter, we will further examine how xenophobia is associated with climate justice and look at some intervention strategies we can use to address the DEI issues in climate justice. We will look at assessment issues and intervention strategies at the micro, mezzo, macro, and eco levels of practice, as illustrated in Figure 8.2.

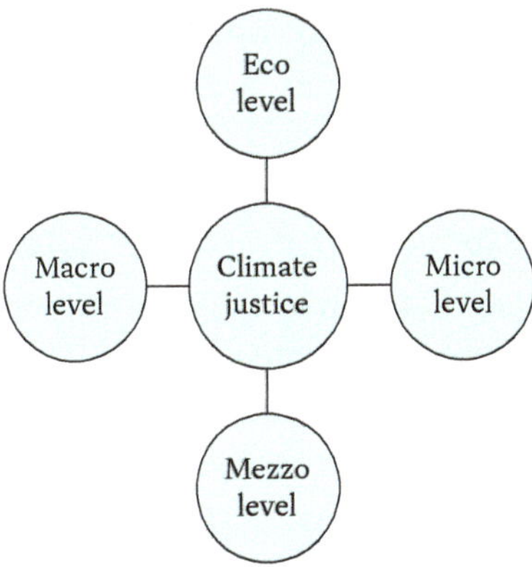

FIGURE 8.2 A Model of Four Levels of Practice With Xenophobia and Climate Justice

Micro-Level Practice With Climate Justice

Xenophobia and Biophobia

When we work with individuals, couples, families, and small groups, we have many opportunities to help people make a positive difference in climate justice. Perhaps one of the most important way any of us can make a positive difference is to become conscious of our own xenophobia and make changes in the ways we think about and behave toward other people. As professional helpers, we all have a "response-ability" to do that work ourselves and then help others do the same.

As illustrated in Figure 8.3, we can teach people how xenophobia and projection are interrelated. Thus, our xenophobia can lead to projection, and vice versa. However, as we become more conscious of our xenophobia, we can also become more aware of our projections, and our increased consciousness of our projections can lead to greater consciousness of our xenophobia.

The world's major wisdom traditions all seem to recognize our xenophobic tendencies, and all teach some form of the golden rule, which instructs us to behave towards others as we want them to behave toward us (ReligiousTolerance.org, 2021). Krishnamurti (1995) writes, "We are all one humanity, caught in different spheres of life." And Vaughan and Walsh (1988) summarize that we make a person a "stranger" when we misperceive them.

Helping professionals can help people heal and transform their xenophobic reactions of others as well as become conscious of our projections so that we do not turn any of our fellow human beings into strangers. As we become more aware, we can start to see the humanity in them and others. Fear decreases, and cooperation across differences is more possible in our families, institutions, and communities. We can start to care about other people who are suffering from the climate crisis and thus be able to help co-create a climate change in our humansystems.

In earlier chapters, we suggested that the ego evolved to help our ancestors survive and, probably later, to justify the accumulation of wealth and power. The ego's tendency to see the self as separate and special probably gave individuals and groups of people a rationalization for their seizure of other people's resources and lives. One of the functions of the ego is to compare the relative value of individuals and groups of people. When I compare myself as either superior or inferior with others, these views are usually associated with my ego.

Evidence of how xenophobia is associated with ego goes back to the start of recorded human history. We see a tendency in many cultures to create hierarchies of superiority and inferiority throughout history that Wilkerson (2020) calls *castes*. Slavery in ancient Greece and Rome, for example, was justified by the belief that those in power were superior to their slaves. These days the term *xenophobia* is used often to describe those who discriminate against foreigners and immigrants, believe their own culture is superior, and seek to keep those they think are undesirable out of their community and country.

In our text, we will define *xenophobia* as including our fear of anyone whom we experience as unfamiliar, different, or otherwise threatening. *Xenophobia* is derived from the Greek roots *xenos* ("stranger") and *phobos* (fear"). As a human, I may fear those who seem to be different from me in any dimension, including race, gender, culture, politics, religion, nationality, sexual orientation, ability, and age. As Robin May Schott (2023) writes, "Xenophobia is a fear of individuals who look or behave differently than those one is accustomed to." *Projection* can be defined as the process of transferring my own feelings, thoughts, or behaviors onto others, including other human beings and other living things.

Our fear of other people can, especially when it remains unconscious, easily lead to distortions in how we see ourselves and the world. This is because we often deny or minimize our xenophobic fears out of our own shame and guilt about them. Perhaps the most common and psychologically "primitive" way our ego might justify xenophobia is by projecting our own negative traits onto the people we fear, such as by thinking that "all those Orange People are lazy, criminal, and unintelligent." Thus, in our xenophobia, we usually project the "shadow" parts of ourselves onto those people we fear, which then justifies our misbehavior toward them.

How can we interrupt this cycle of increasing fear, projection, and violence? One way in which helping professionals can intervene is by focusing on consciousness work with both xenophobia and projection. As Figure 8.3 illustrates, xenophobia and projection are interrelated processes. As I become more conscious of xenophobia, it has less power in my own psyche, my fear can diminish, and I can start to see myself and other people more realistically and with acceptance. As I become more aware of my projections, the same things happen; I can become more cognizant and accepting of my shadow aspects (limitations) and therefore will be less likely to project my shadow parts onto others.

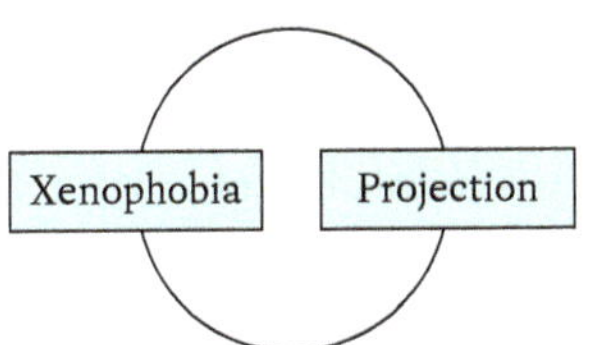

FIGURE 8.3 Xenophobia and Projection Are Interconnected

We can also strive to help people by encouraging both new ways of thinking and of acting. For example, sometimes we justify our xenophobia by telling ourselves that what we could call *xenophilia*, or the love of strangers, is actually more dangerous than xenophobia. An example of this thinking happened in my own community years ago. When asked, some leaders on the more affluent and White east side of town refused to invite teen basketball players from the other west side of town to join their basketball league. Sadly, they did not see the bigger picture—that when we get our children together to play sports or interact in other positive ways, we can increase understanding and reduce conflict and violence between populations. One way to help such a culture of fear to transform is to set up opportunities for the adults in the community to begin meeting with each other, perhaps in neighborhood dialogues or through their churches, community markets, and gardening projects.

How else can we help clients deal with their fears of xenophobia and biophobia more adaptively? I have found that my clients and students respond best to a combination of consciousness and behavioral change. For example, I often recommend Susan Jeffers's (1997) *Feel the Fear ... and Do It Anyway*, which offers a readable and concise approach that most people like. Her idea of "feeling the fear" is analogous to our idea of reverent awareness, and "do it anyway" suggests the common sense idea of not letting fear inhibit the behaviors I want to express. For example, I might, after becoming conscious of my fear, decide to go ahead and talk with someone at the office or in church who belongs to a different ethnicity or has a different political view from mine.

We may all live in a period in which xenophobia is increasing. There is evidence now that in the United States, we more commonly avoid both other people and nature in our daily lives and that these fears of xenophobia and biophobia may actually be associated with each other (Seo, 2023). Helping professionals may want to respond to this increasing fear by naming it and promoting more positive human interactions.

Climate Justice and the Radical Middle

As in all of inclusive practice, our work in healing xenophobia and supporting climate justice is *radical middle* work. We neither deny the differences between people nor amplify them, but rather support our clients, students, patients, and communities in becoming conscious of how things actually are and how both similarities and differences between people always exist.

For example, mental health experts continue to debate over whether xenophobia should be considered a mental health disorder (Fritscher, 2021), and regardless of where we stand on that issue, we helping professionals can address xenophobia as a distortion of reality that can cause harm to ourselves and others.

This is because when I dehumanize another person or group, I cannot help but also dehumanize myself, as I give in to fear and lose touch with the love I have in my heart. There may be circumstances in which it is helpful to characterize the distortions of reality involved in xenophobia and projection as related to the criteria we use today for some of our existing mental health disorders.

We want to always have compassion for people, which does not mean that we condone their actions; however, it does mean that we recognize that we all are human, and, therefore, we all have both strengths and limitations. We realize that our often-dominant left-brain functions tend to place people into such categories as race, gender, culture, politics, religion, nationality, sexual orientation, ability, and age (McGilchrist, 2009). My left-hemispheric functions tend to assign each person the stereotypical traits that I believe belong to the category to which my brain has assigned them. This categorization happens literally in seconds, when I first meet someone, and the biases and projections that occur in those moments tend to persist. Such first impressions are quick, common, but, fortunately, also reversible (Mann & Ferguson, 2015). Helping professionals can assist people in becoming conscious of these left-brain tendencies so that we do not have to necessarily trust all our first impressions.

We also encourage people to let their hearts break open if they can. We realize xenophobia and biophobia are fears and that when we act out of fear, we tend to have poor outcomes. Krishnamurti (1996) taught that fear ultimately comes from our desire for security, permanence, and certainty. We understand that it is OK to want security, permanence, and certainty; however, we also realize that such comforting experiences rarely exist for long on Earth. We help our clients see that as we become conscious of our fears and accept the unpredictable nature of the world, we become more capable of loving ourselves for who we are and the world for the way it actually is.

Ultimately, we helping professionals can help people see that love-based behaviors tend to bring better outcomes than fear-based behaviors. We can talk with our clients about fear and love and help them become more conscious of their inner and outer worlds. As consciousness increases, people can become more compassionate toward themselves and others who are different from them. Thus, our consciousness is associated with our own well-being as well as with that of other people, other livings, and the ecosystems that support all life.

Climate Justice Work, Civility, and Inclusivity

In my DEI work over the past decades, I have sometimes observed a lack of civility and inclusivity within DEI committees, task forces, classrooms, and communities. I have seen both professionals and members of the public sometimes monopolize conversations and ignore the voices of other community members with less institutional power or social status. I have watched as members of committees occasionally made cruel verbal attacks on their fellow members when they used pronouns or other language that they saw as inappropriate. Sometimes participants also angrily scold others for having views different from their own regarding such things as philosophy of change, beliefs about DEI, or personal opinions about local or national news events.

In my view, we cannot foster inclusivity by being exclusive. Uncivil and unwelcoming DEI committees cannot effectively help co-create a civil, inclusive, and just society. Perhaps such misbehavior can be understood as understandable trauma responses (i.e., responses to past traumatic events) that are being misdirected toward colleagues in the present moment. On the one hand, we cannot expect people to always be able to leave their egos and trauma at the door before meetings. On the other hand, we could, together, help co-create more inclusive communities in which we strive to follow the dialogue ground rules we have discussed in this text. One of the most significant rules in this regard is to offer amnesty. We all make mistakes, and we all inevitably will offend each other—not because we evil or insensitive, but because we are humans.

As a chair of DEI meetings, I have learned to use some of the following principles to facilitate meetings:

1. Ask participants to follow the dialogue ground rules.
2. Ask (not demand) that those with the least power speak first, if they wish to.
3. Allow for time for people to give each other feedback in a civil manner.
4. Encourage sharing but discourage performative contests regarding who has suffered the most or is the "most enlightened" or "least xenophobic" member.
5. Share leadership with everyone as much as possible.

Grandiosity, Disconnection, Power, Wealth, and Fame

Although we have discussed the healing and transformation of such ego-related traits as grandiosity, disconnection, and the desire for power, wealth, and fame, this topic is worth reexamining again briefly in this chapter on climate justice. We basically want to become conscious of these traits inside ourselves so that we become less likely to act them out in ways that are unconscious and harmful to ourselves and others.

I have found that most people in our culture today tend to take one of two extreme positions regarding our egos. This idea is illustrated in Figure 8.4. As you can see, we have, on the left side of the continuum, the extreme denial position: "I don't have an ego." This position is perhaps represented by today's archetype of the "nice person" who carries guilt and shame about their own ego and tries to hide from sight and keep it in the shadows. On the right side is the other extreme: "I have an ego but don't take responsibility for my behavior." This position is perhaps best represented by today's archetype of the political strongperson, who attacks those by whom they are threatened and publicly celebrates their grandiosity, narcissism, and greediness.

As you can see, the radical middle position is: "I have an ego and take responsibility for my behaviors." In this case, the person is conscious of their ego and owns any egoic behavior that they display. Such reverent awareness allows me to heal (i.e., accept) myself as I am and transform my behaviors so that I can

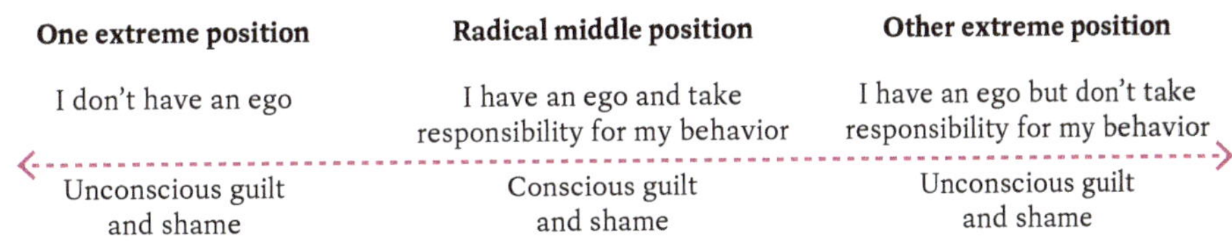

FIGURE 8.4 A Radical Middle Position Between Two Extreme Positions on the Ego

become free of egoic behavior. In other words, when I am able to see myself without any trying to change what I see, the ego no longer has power.

We can see that the two extremes illustrated in Figure 8.4 are actually two sides of the same coin, a characteristic often seen in the two extremes of most continua. Both extremes are associated with *unconscious* guilt and shame. The unconsciousness of the nice person archetype results in the attempt to minimize or deny ego. In contrast, the unconsciousness of the political strongperson archetype results in the attempt to seek public approval for ego by pretending that xenophobia is a strength rather than what is actually is—a fear-based weakness. However, as Figure 8.4 suggests, when we make guilt and shame conscious, as seen in the radical middle position, healing and transformation are possible.

In Section 1, we studied how our own consciousness is perhaps the best medicine for transforming our egos. We can also see this approach described in many world wisdom traditions. As helping professionals, we can offer mindfulness and meditation classes that are not just about creating more inner peace but also about the healing and transformation of ego (two goals that go well together). Some of our community members may find that a spiritual or religious practice will also help them with this work, and we may refer them to classes that offer meditation, yoga, prayer, or mindfulness instruction.

Healing and Transforming Guilt and Shame

When I have co-facilitated DEI workshops for groups of White people, there are often would-be participants—often involuntary participants—who would complain that we were trying to make them "feel bad about ourselves and our country." I've learned to thank the person for their comment, congratulate them especially for being willing to own their feelings of guilt and shame, and then say something like, "If I thought it would help our country and our world to make people feel even more guilt and shame, I might consider trying to do that. But it doesn't look to me that guilt and shame helps. In fact, most people I know already live with a tremendous amount of guilt and shame every day, and it does not appear to have done much good yet."

Most likely, these participants probably walked into the workshop already full of unconscious guilt and shame and wanted to blame the workshop for these uncomfortable feelings. Actually, we helping professionals generally want to support people in healing and transforming their unconscious guilt and shame, because these fear-based feelings typically do not actually help them bring more loving kindness toward themselves and into the world.

We also want to recognize that, although they are related, guilt and shame are also different. *Guilt* is negativity about what I did, whereas *shame* is negativity about who I am. In Western culture, guilt is a *fear* of punishment, whereas shame is about feeling not enough and therefore *deserving* of punishment and suffering and *not deserving* of peace and love. In my experience, most people feel both guilt and shame about how their egoic functions think and behave as well as about their general lack of connection with the world (as will be discussed further in Chapter 15).

When we attack another person verbally or physically, we may initially feel a moment of what the Germans call *schadenfreude*, which is egoic happiness about the suffering of another. However, most of us probably feel bad later, as we rise above ego and access more of our brains and hearts. In our culture, although most of us experience plenty of guilt and shame, we tend to try to push these feelings away and keep them hidden from ourselves and others. When feelings remain unconscious, they cannot be healed or transformed.

Feeling bad about ourselves usually fails to keep us from causing harm, but it often does lead to self-harm and harm to other people, other living things, and ecosystems. It is my consciousness, my reverent awareness, which helps me love myself and my world. Guilt and shame heal when we become conscious and accepting of these internal experiences. By becoming conscious of my guilt and shame, I also gain the ability to see with clarity why I experience those reactions and how my repression of my guilt and shame makes it only more likely that I act out my egoic traits in harmful ways. Ultimately, as I learn to accept that we all have limitations and make mistakes, I am less likely to be haunted by guilt and shame.

Addressing All the Ecobiopsychosocialspiritual Dimensions

Throughout the text, we have looked at how inclusive practice addresses all the ecobiopsychosocialspiritual dimensions in our casework. For example, in the psychosocial dimension, since we know that most people benefit from taking positive actions in response to crisis, we can refer people to organizations that give them opportunities to socialize with other committed people with whom they can receive information and take actions that make a difference. The Climate Justice Alliance (2023) is one such organization that can have psychosocial benefits for participants; students, clients, and patients can find more information on the organization's website and opportunities to volunteer or otherwise get involved in helping to address climate justice issues.

As we will further explore in Section 3, we can work with people in the biological dimension. As we help them get back in touch with their own bodies, many people find that they begin to open up more to their connection with nature. (For more information on the definitions of what *nature* is, please look at the final section of this chapter.)

The spiritual dimension is probably that with which professional helpers are generally the least comfortable. In the literature, *spirituality* is generally defined as *individual* dimension of human development, in which people find a sense of connection and purpose in the world. In contrast, *religiosity* is defined as a *social* dimension in which people share beliefs, rituals, and doctrines within their community (Derezotes, 2005).

People working within faith systems have often led the way in transforming the human tendency toward xenophobia. In the 1950s and 1960s, for example, Dr. Martin Luther King, Jr., was joined by many other U.S. Christians in his civil rights movement and many African American churches across the South, as they served as meeting locations for activists and citizens to support the antiracism movement.

More recently, in response to attacks on foreign nationals in South Africa, United Methodist Church (UMC) congregations united in a day of prayer to protest against xenophobia. Reverend Mills Maliwa stated that the UMC is against *xenophobia*, which is defined as any fear and hatred of foreigners or strangers (Mkwalo, 2019).

Most spiritual and religious traditions recognize the wisdom of the golden rule (ReligiousTolerance.org, 2021). Our expressions of love can range from a superficial level of politeness to a much deeper consciousness about humanity's emotions, beliefs, biases, judgments, and behaviors. One point often missed is that since the golden rule means loving you as I love myself, then self-love is essential. G. I. Gurdjieff addressed this point when he encouraged students to stay present "to the discontinuity between my wish to be able to fulfill this commandment and my personal incapacity to do so, due to the many attitudes and prejudices conditioned into my psychology by my surrounding social structure" (Aronson, 2021, p. 13). By doing so, the student could learn compassion for the humanness of others by first learning compassion for their own strengths and limitations.

Mezzo-Level Practice With Climate Justice

Just like there is systemic racism and systemic sexism, we can say that there is also systemic climate injustice. *Systemic injustice* may be thought of as the increased exposure of one population to negative climate crisis–related impacts perpetuated by prevailing inequities within our humansystems. Examples of populations who are especially vulnerable to such systemic climate injustice issues are listed in Table 8.1. Intersectionality in systemic climate injustice occurs when populations at risk have multiple identities that make them even more vulnerable to climate crisis–related impacts: For example, a person who identifies as being both part of the LGBTQIA+ community and part of a racial minority and who lives in a poor country involved in an ongoing civil war is especially vulnerable to such weather events as drought or fire or flooding. This is because in part, they are less likely to have a strong preexisting support system and material resources to assist them in times of crisis.

Dialogue can be a useful tool in promoting climate change transformations in our mezzo-level humansystems, including systemic climate injustice and related issues in intersectionality. As we discussed in Chapter 6, researchers have found that participants in intergroup dialogue often report positive changes in the way they feel, think, and act toward people in other groups.

Intergroup dialogues may be especially helpful in fostering mutual understanding and cooperation about the climate crisis in local communities. In these conversations, the facilitators invite members of at least two groups of people who have differences that currently divide them; for example, perhaps a group of climate change deniers and a group of climate change activists might both be invited. It is important that approximately equal numbers of people in each of two groups are attending so that a relatively safe dialogue can occur (no dialogue, of course, can ever be completely safe for people, since people can easily feel emotionally vulnerable and sometimes be attacked during any meeting).

With groups who are meeting for the first time, are composed of people who tend to dehumanize each other, and may not have never participated in a dialogue, the dialogue ground rules can especially help provide a foundation for progress in relationship building. The effective facilitator uses the ground rules as a teaching tool and expects that most people will accidently violate the rules at times, since that is what most of us do in our everyday lives. As long as people make a genuine effort to follow the guidelines, we will continue to support their participation in the group and offer amnesty.

It is also best to try to create a series of ongoing dialogue meetings, rather than a single event, since many people need some time to learn to start to trust each other before they are willing to go deeper into conversations. Facilitators can ask participants to own their own biases, projections, and judgments in the meetings, and when this begins to happen, there can be real healing and transformation. Most people are typically unwilling to speak about these kinds of beliefs, except perhaps when they are angry, and when we share our biases in anger, we end up only attacking and further alienating each other. In these dialogues, the purpose of such sharing is to heal xenophobia and transform our ways of thinking, feeling, and acting. Xenophobia heals when we become conscious of it, and it is transformed when we change the way we act in the world. It may take some participants multiple sessions before they are ready to do such intimate sharing.

Macro-Level Practice With Climate Justice

In this section on macro-level practice, we will explore some of the suffering to which some at-risk populations currently are being exposed and some macro-level interventions that helping professionals can use in response.

A great place to start in our reflection on macro-level climate justice is to reflect on the words of Naomi Klein:

> The accelerated need for growth has made our lives more precarious. ... Elites who benefit greatly from these [growth] priorities are the same ones who bankroll political and media projects devoted to pitting nonrich people against one another based on race, ethnicity, and gender expression—making them less likely to unite based on common economic and class interests. (Klein, 2023, pp. 228–229)

We helping professionals can help our communities redirect their focus away from the so-called culture wars and instead look at ways to hold the major polluters of the world accountable.

Who are these elites who continue to benefit economically as they contribute a large proportion of GHGs still going into our atmosphere? A recent news briefing from the Carbon Majors Database reports that:

> 57 industrial and state entities have since the 2015 Paris agreement produced the lion's share of GHG emissions. About one-third are investor-owned oil companies like the United States–based ExxonMobil and Chevron and England's BP. State-owned energy corporations such as Saudi Aramco and Russia's Gazprom make up another 36%, and nation-state producers, such as those in China and Russia, comprise the final 31%. Overall, the top 57 main offenders account for a whopping 88% of global emissions, comprising a shocking 251 gigatons of carbon dioxide. (Al-Sibai, 2024)

Such research findings can be disseminated to the public, who, as consumers and voters, can express our opinions through our purchases and votes.

We can also disseminate information on who the climate crisis continues to harm the most. For example, Jeff Goodell (2023) has reported that the people in the United States who are most likely to be harmed by excess heat are unsheltered people, poor people, and agricultural and construction workers. Most farmhands and construction workers in Texas are currently migrants from Mexico and Central America. Policies can be implemented to help protect vulnerable outdoor workers. However, unfortunately, other policies are also being put into effect that make our efforts to protect workers even more difficult. For example, Texas Governor Greg Abbott signed legislation in 2023 that prohibits any city or county in Texas from passing laws that expect protective shade and water breaks for outdoor workers (Goodell, 2023).

At COP28, International Rescue Committee President and Chief Executive Officer David Miliband has reported that our climate crisis has increased the poverty of the vulnerable poor in many locations across the world (CNN, 2023). This is because those populations can often not afford to move to the safest and most desirable living locations, nor can they pay for losses from fires and floods.

However, there is evidence that the wealthy are gradually becoming more vulnerable as well. Recent research suggests that although more affluent populations still have more protective green growth in their neighborhoods, the "luxury" shield of wealth no longer serves to keep wealthy urban neighborhoods significantly cooler than poor neighborhoods. This is, in part, because extreme heat waves are now reducing the green growth in *all* neighborhoods (Rainey, 2023).

In *Climate Resilience for an Aging Nation*, Danielle Arigoni (2023) reports that climate disasters such as heat waves, fires, and flooding disproportionately affect the aging. She argues that the protection of aging populations should be prioritized through planning policies in transportation, emergency response, housing, and health care. Other populations, including many disabled people, may be equally vulnerable because they may have similar physical and mental restrictions as the aging population.

President Biden signed an executive order (EO) in April 2023 that created an Office of Environmental Justice to help reduce pollution's impact on communities of color (Walsh, 2023). Biden built on former President William Clinton's 1994 EO that required federal agencies to include environmental justice in their missions. Although these changes are certainly positive news for environmental justice, there is still much work to be done. Indeed, the 1994 EO directed agencies to adopt an environmental justice strategy and then implement it, but not every federal agency fulfilled this mandate because of various enforcement constraints.

Resource availability is not distributed equally across the globe, and some populations use a disproportionately large amount of resources. For example, the United States has the highest water consumption per capita (at 7,800 liters a day) among the 25 most populated countries, and six large counties in the world consume about half of global water supplies, with China using 16%, India 13%, the United States 10%, Brazil 4%, Russia 3%, and Indonesia 3% (Winters, 2023). Professional helpers can offer such information to people in their community so that they can make more informed decisions about whom they elect to public office.

Coal miners are an example of a population that is especially vulnerable to losing income and wealth because of efforts to reduce GHG emissions. Wind farms, which are often suggested as a replacement for coal burning, may employ only about three people in each farm (Hinde, 2023). Populations in regions such as West Virginia that depend upon coal mining for their income are understandably least enthusiastic about wind farming. Policy makers can create plans that support such populations so that they can receive help, such as job training, assistance with relocation, and income subsidies.

There are about 3 million people in the United States who have an opioid use disorder or are dependent on heroin; these people make up another population at risk from the effects of the climate crisis. According to researchers:

> Cities with high overdose rates ... are also at the highest risk of extreme heat, drought and flooding, and the opioid epidemic is likely to peak just as climate change's most brutal effects start to take root. ... During environmental crisis, drug users frequently experience more social and economic hardships than the rest of the population. These hardships often snowball into poor health outcomes, including death. Climate change will likely disrupt the illegal drug supply chain—making dangerous drugs even more variable and deadly. Soaring temperatures will make people more vulnerable to overdose. Despite this, drug users are typically not considered in public health efforts related to climate change preparedness. (Ezell, 2023, pp. 14–15)

Researchers can now track which regions are most at risk due to the climate crisis. For example, a review of a climate crises map shows where the neighborhoods that are most vulnerable to extreme weather events such as fire, floods, and pollution are located in the United States. The top 10 counties on the map are located in the South, and half of the 10 are in Louisiana (Yoder-Grist, 2023). Our communities can support the work of scientists who work in these regions to uncover why they are so vulnerable and create appropriate prevention and emergency response plans.

Fair Trade

What is *fair trade*? Fairtrade International (n.d.) explains, "Fairtrade [sic] changes the way trade works through better prices, decent working conditions and a fairer deal for farmers and workers in developing countries [that] enables farmers and workers to have more control over their lives and decide how to invest in their future" (p. 1).

D'Francisco VanderHoff Boersma (2010) associates the need for fair trade with current Western economic philosophy:

> Today, neoliberalism and its holy Trinity—deregulation, innovation, and globalization—are facing a crisis, and we are finding out that the trendy notion of "sustainable development" is, for all practical purposes, an oxymoron. The time is perhaps ripe for to rethink our ways of doing things and fight the spread of individualism and consumerism, even if it is now always easy to go against the current. (p. 6)

How do fair trade programs support climate and economic justice? First, fair trade reconnects us with our environment. Many of us living in wealthy countries do not know where our food and other products come from and are out of touch with the often terrible working conditions that poor farmers and factory workers suffer from every day. The origins and destinations of some of the most common imported food and products we enjoy and consume daily are described in Table 8.2, in which we list products and information highlighted in Éric St-Pierre's *Fair Trade: A Human Journey* (2010).

Second, the fair trade premiums that wealthy consumers now pay for products from poor countries, which often include additional organic premiums as well, help raise the standard of living for poor workers and help them protect the environment in which they live. Examples of these benefits from Fairtrade International (2018) are shown in Table 8.3.

TABLE 8.2 Some Products From Developing Countries That Need Improved Fair Trade Protections

Product	Notes
Handicrafts	Efforts are being made to improve market sales and prices for handicrafts
Coffee	Main producing countries: Brazil, Vietnam, Colombia, Indonesia, Ethiopia
Cocoa	Main producing countries: Côte d'Ivoire, Ghana, Indonesia Nigeria, Brazil
Sugar	Main producing countries: Brazil, India, China, Thailand, Pakistan, Costa Rica
Tea	We drink 25,000 cups per second; global trade is approaching $4 billion
Flowers	Successful fair trade efforts show that clients in the North are willing to pay more
Rice	Main importing countries: Switzerland, France, United Kingdom, Germany, Italy
Cotton	Main importing countries: United Kingdom, France, Switzerland, Germany, Finland
Bananas	Main importing countries: United Kingdom, Switzerland, Germany, United States, Austria
Wine	Fair trade significantly reduced economic injustices in Argentina, Chile, South Africa

TABLE 8.3 Eight Ways Fair Trade Farmers Help Protect the Environment

Eight Ways	Notes
Reforestation	Farmers often invest their fair trade premium in reforestation
Organic farming	Fair trade forbids use of many harmful agrochemicals
Tree planting	Many farmers grow crops and trees together to increase yields
Climate emergency	Farmers trained how to switch to new green methods when necessary
Wildlife conservation	Many farmers use their premiums to protect wildlife and environment
Carbon credits	Companies buy credits for the emissions they produce.
Green energy fuels	Fair trade premiums help farmers switch to green energy fuels
Reduce water usage	Premiums also help farmers pay for projects that reduce water usage

Eco-Level Practice With Climate Justice

What Is *Nature*?

Before we explore eco-level practice, we might want to ask, what, in fact, is *nature*? The answer is, we don't all yet agree on a definition. The English word *nature* was formed in the 13th century from the Latin word *nascor*, which means "to get born." Obviously, our use of the word has changed, broadened, and become more complex since that time and, more recently, has become part of our complex scientific, cultural, philosophical, and political language.

Probably my favorite definition of the word *nature* comes from the Encyclopedia of the Environment (2021), which explains how the concept of *nature* has evolved across time and locations and that no single definition is universally accepted. We also learn that *nature* has been used as a casual term in European languages and has rarely been addressed by any advanced academic theorization, except when used in the phrase *human nature*. The encyclopedia offers four definitions, listed in Table 8.4.

TABLE 8.4 Four Definitions of Nature

(1) The totality of **material reality that does not result from human will** (as opposed to artifice, intention and culture)

(2) **The whole universe** as a place, source, and result of material phenomena, including humans or at least the human body (as opposed to the supernatural, metaphysical, or unreal)

(3) **The force at the core of life and change** (opposing inertia, fixity, and entropy)

(4) The **essence**, the set of specific physical properties and qualities of an object, living or inert (opposing denaturation)

Source: Encyclopedia of the Environment, 2021 (bold segments are by authors of article).

As the encyclopedia suggests, the fact that we have multiple definitions of *nature* makes it important to agree on what definition or definitions we are referring to when we use the term. I find that there are useful elements in all four definitions.

In eco-level practice, when I think about the attitude I want to have and model toward nature, I think about having a reverence for all living things (including all humans), for the ecosystems that support life, and for the basic elements and forces that underlie our universe. And when I think about protecting nature, I want to carefully consider at least the five questions that are outlined in Table 8.5.

TABLE 8.5 Five Initial Questions to Consider When Seeking to "Protect Nature"

(1) What do we want to protect?
(2) Why does it need protection?
(3) What do we need to do to protect it?
(4) What are the known potential impacts of such "protection" (and of lack of any protection) on the larger ecosystem?
(5) With which human individuals and populations do we need to cooperate to get this done?

Biophobia

Perhaps one of the most basic and most common fear we humans have, across the world, is our fear of nature, or *biophobia*. We speculated that biophobia may have roots in the lives of our ancient ancestors, who had fear of dangerous animals and weather. Psychologists believe our fear of the dark developed because our ancestors had to stay vigilant at night to avoid nocturnal predatory animals (Norton & Antony, 2021).

I have heard many of my students, clients, and friends talk about how scary it can be outside in the dark at night, either in a wilderness location or on a city street. When people experience this fear, we helping professionals can reframe the alarm and sometimes trauma that folx often have with nature. We can suggest to them that this fear is an ancient emotion, perhaps largely passed on through our genetics, that can remind us of our deep connection with nature.

Many of my social work students in my spirituality in social work classes have, over the years, elected to do a vision quest in which they might spend two days and a night out in nature by themselves. They usually take a journal and camera and camping equipment and reflect on what they are feelings and thinking. Often the journals they write have entries about their biophilia and how they started to overcome their fears of nature during their trip.

Redefining Sustainability From the Perspective of Traditional Ecological Knowledge

In *Beaver, Bison, Horse: The Traditional Knowledge and Ecology of the Northern Great Plains*, Grace Morgan (2020) writes about traditional ecological knowledge:

> Ecosystems are not just assemblages of organisms: they are expressions of human, floral, and faunal co-evolution in response to past environmental conditions. They are shaped by Traditional

> Ecological Knowledge (TEK): knowledge and practices passed as stories and songs from generation to generation informed by strong cultural practices and rituals, such as traditional dances. TEK includes sensitivity to change and reciprocity, that is, a give and take relationship with nature that acknowledges a responsibility to care for and respect the rights of all living beings. ... Unlike Western science, TEK observations are qualitative and long term and are about contexts and relationships ... and inseparable from a culture's spiritual and social fabric. (pp. 188–189)

Morgan credits anthropologist Gregory Cajete with the invention of the term *ecosophy* to describe a Native American science that is essentially an epistemological way of living in the world.

TEK can help inform our eco-level practice across all local and global cultures and races in our world. I love the core idea that Morgan (2020) credits to Anishinaabe ecologist Robin Wall Kimmerer, that "it takes all of what it means to be human—body, mind, and heart, and spirit—to understand something ecologically" (p. 189). Morgan goes on to write about Tohono O'odham ecologist Dennis Martinez' redefinition of *sustainability* as a "kincentric" view of nature that respectfully views all life as our kin and the Earth as our nurturing mother. In Chapter 12, we will further examine how we humans can learn to relate from a more kincentric position with all of nature by investigating our own biopsychosocialspiritual relationships with the elements of earth, wind, fire, and water.

When we work with our clients, students, patients, and families, we can not only teach ecological knowledge (e.g., statistics about the climate crisis) but also offer a more inclusive cultural framework that emphasizes the values of a "Kincentric Ecology" (Martinez, 2019). The idea of kinship with nature potentially reconnects us with the world in which we live, as we see other living things and ecosystems not as things to control and conquer but rather as our actual relatives whom we love and respect.

I have found that most people across both cultures and the lifespan can understand and express this shift toward kinship ecology. In my work with children and adolescents, for example, young people quickly grasp the idea of kinship, at the cognitive, emotional, and spiritual levels. At the other end of the lifespan, I have found that most of my current Osher students are equally open to the value of kinship and eager to explore its implications in their daily lives. These classes are part of the Osher Lifelong Learning Institutes in the United States, which generally provide noncredit courses to people older than age 50.

What Does the Land Want? What Do Other Living Things Want?

Our eco-level climate justice work requires that we expand our ways of knowing from the logical and scientific left-hemispheric brain functions to also include the right-hemispheric brain functions of imagination and intuition. As neuroscientist Iain McGilchrist (2009) has described, humans being have multiple ways of knowing that can help inform our thinking and actions.

I love the question that Yomi Wrong (2023, p. 8) asks in *Orion Magazine*: "What does the land want? We gather in ceremony to hold this central question, trusting the forest will answer in its own time." We may find in the decades ahead that all four ways of knowing will contribute to the answer to Wrong's question.

Wrong's article goes on to advocate for a new view of the relationship of disabled people to the environment:

> We have been told a terrible, violent lie that disability is incompatible with nature, that accessibility is antithetical to preservation. This view has severed many disabled people's relationship

> to wilderness, rendered us marginalized in matters concerning the environment, and keeps us sidelined in liberatory movements for land rematriation and restoration. (2023, p. 8)

Perhaps we can expand this idea to include all populations—nature belongs to us all.

A search for the meaning of the word *rematriation* helps us distinguish it from *repatriation*, which is a return to the place where I was born, whereas *rematriation* refers to a return to a life full of reverence for nature. One source provides the following definition: "Rematriation is Indigenous women-led work to restore sacred relationships between Indigenous people and our ancestral land, honoring our matrilineal societies, and in opposition of patriarchal violence and dynamics" (Sogorea Te' Land Trust, 2023, p. 1).

The agony of the Earth is perhaps best felt in the human heart. I have often felt myself feeling sad when I see the wild land in Utah scarred by the activities of people, the water of a mountain stream full of bubbling pollutants, or the mountain valley air polluted with chemicals from industrial and transportation industries.

El Akkad (2023) asks:

> What does going back to normal mean when normal is the one thing that's never coming back ... that sheer variety of potential disasters—the many ways in which life in the Anthropocene has become deeply unpredictable—has led to more uncomfortable questions, such as, At what point does a place become so susceptible to burning or flooding or storms that its residents have to start thinking about the viability of remaining there ... the idea of adaptation to the environment rather than control of it is till for many people, a tough sell. (p. 21)

In addition to listening to the land, water, and air, humans can listen to other living things. Since many people today still doubt that any nonhuman lifeform can have thoughts or feelings, we can turn to scientific evidence that supports the idea that other living things do, in fact, think and feel. For example, Mark Bekoff (2020) wrote an article for *Psychology Today* in reaction to the efforts of some Canadian authorities to deny that animals think and feel. He states:

> The Ontario Federation of Agriculture's conceptualization of the cognitive and emotional lives of clearly sentient beings is pure fiction and should be read as such. ... How we treat these and other clearly sentient nonhumans isn't necessarily a matter of rights. Rather, it's a matter of decency and depends on using what we know—and have known for a long time—on the animals' behalf. Indeed, we are obligated to do so. (pp. 1–2)

Bekoff goes on to summarize that although scientists have found evidence of the cognitive and emotional lives of nonhuman animals, the Federation of Agriculture had motives to hide the truth in order to support their own agricultural agenda (i.e., in support of raising farm and ranch animals).

Bekoff (2020) adds:

> People who know anything about the field of cognitive ethology (the comparative study of animal minds and what's in them) pay careful attention to what other animals know and feel, capacities and adaptations that allow them to be card-carrying members of their species, not ours (or that of other nonhumans). Intelligence is a slippery concept and should not be used to assess suffering. Asking if chickens suffer less than pigs, or if pigs are as smart as dogs, is meaningless and idle speciesism. (pp. 2–3)

Those that doubt that other animals can be sentient are often even more skeptical about the cognitive or emotional lives of plants. However, David Kuchta (2023, p. 1) reports that "animals like us have memories that allow us to avoid future pain. Memories of previous injuries don't trigger our bodies to produce adrenaline when the memory of that event returns. ... Without a central nervous system, plants don't process sensations that way. But plants do react to negative or unpleasant sensations." Scientists have discovered that many plants, just like many animals, can communicate with each other, using sounds as well as chemicals.

In a wonderful book I found on honey bees, author David Camp (2020) writes about research he and others have done to try to find out whether bees show or feel emotions or become depressed. They found that pessimistic bees tend to treat uncertain stimuli like a punishment, just as depressed humans might experience hostility when they encounter a neutral gaze from another person. "Honey bees have become the first invertebrates found to exhibit pessimism—a benchmark cognitive trait supposedly limited to 'higher' animals. ... In short, the bees acted as if they felt pessimistic, and their brains reflected this" (p. 84). At a time when insect populations around the world are dropping and threatening the health of ecosystems (and human farming systems) that depend on insects for their well-being, this finding is significant. Has our climate crisis had an impact on not only the physical but also the psychosocial well-being of all living things on Earth?

I often think that any human who has an open mind, heart, and soul can sense the suffering of an animal in distress. I have also had similar reactions to plants, such as trees that have been hit by lightning. I remember many examples of animal suffering that I have seen, including a deer who was just hit by a car, a bird just shot by a hunter, and a fish just pulled out of the water with a hook in its mouth. We do not write these words to try to make people feel guilty about driving, hunting, or fishing. However, perhaps we can teach ourselves and then our children to at least hold reverence for the animals and plants that we kill, either accidently, for sport or for food.

Dialogue conversations, as we will explore further in Section 3, can be created between humans and humans, as well as between humans and the rest of the universe. In all forms of dialogue, the most important skill is listening. Helping professionals can help our students, clients, and patients listen to the land, water, and air and to other living things so that they can help give them all a voice in our own human civilization.

Survivance: Survival With an Attitude

Finally, I was also moved by an article in *Archeology Magazine* by Matt Stirn (2024) in which he reports how ancient Indigenous foragers in what we now call the U.S. Great Basin survived a 1,000-year-long "megadrought" by relocating temporarily from their homelands in the now hot and dry valley floors to the cooler and more productive highlands of what we now call Nevada's *Toquima Mountains*. Stern quotes archeologist David Hurst Thomas from the American Museum of Natural History, who studied the ancient prehistoric village of Alta Toquima, located at an 11,000-foot altitude. Thomas builds on the term *survivance*, which was invented by Chippewa scholar Gerald Vizenor:

> The idea of survival suggests that in a bad situation, people put their heads together and try to get through it on a day-to-day existence. Survivance, on the other hand, is survival with an attitude. No matter how bad the situation gets, people persevere to maintain their culture and connection to important places so that one day, they can turn. (p. 41)

Perhaps we helping professionals, in our climate justice work, can honor the "survivance" of our ancestors who were also challenged by climate change emergencies by teaching and adopting a now-global "survival strategy with an attitude" like they did, many centuries ago. We may *all* need to make sacrifices in order to get through the climate crisis that this time humanity has created, those that are strengthened by our faith-based activist belief that, ultimately, this crisis, too, shall pass.

QUESTIONS FOR REFLECTION

1. What is *xenophobia*? What is *xenophilia*? Can you see xenophobia and xenophilia in yourself and other people? Explain.
2. Can you relate to the idea that there are castes in the United States? If so, in which caste did you grow up, and to which caste do you belong now? Explain.
3. In what ways do you carry guilt and shame about your own xenophobia?
4. Look at the following chart and fill out at least one example for each form of xenophobia listed in column 1 in column 2. When you can, please enter a "price" that the dominant castes may pay for their xenophobic thinking and behaviors. For example, under the category of "Race," I might enter that many White people who have not become conscious of our xenophobia probably experience increased fear and dislike of people of color, which then further intensifies our xenophobia, guilt, and shame. In other words, our unexamined fear of people of different ethnicities leads to beliefs and actions that make us feel even more fearful and ashamed.

Column One Form of Xenophobia	Column Two What is an example of a belief or behavior I have expressed, related to this category of xenophobia?
Race	
Gender	
Queer identified	
Disability	
Religiosity	
Age	
Rank	
Other	

5. Do you see biophobia in yourself? Please explain.

6. Which level of practice do you feel most interested in or comfortable with: micro, mezzo, macro, or eco? Please explain.
7. What is your reaction to the concept of listening to the land? Have you ever felt like you could sense the thoughts or feelings of other living things? Please explain.
8. What is your attitude about paying more for fair trade products? After reading about fair trade, did your attitude change at all? Readers may want to read the beautiful book by Éric St-Pierre that is cited in the text. Although it was published back in 2010, the facts about fair trade are still relevant, and there are wonderful photos that provide us with unforgettable images of some of the people who grow the food we eat and of the environments in which they live.

REFERENCES

Al-Sibai, N. (2024). *Scientists discover the villains destroying the planet: 'Don't blame consumers.'* Futurism.com, April 6, 2024. https://futurism.com/the-byte/climate-change-villains-list

Arigoni, D. (2023). *Climate resilience for an aging nation*. Island Press.

Aronson, A. (2021). The golden rule and the transformation of being. *Parabola*, 10–13.

Bekoff, M. (2020). Do animals think or feel? The Ontario Federation of Agriculture says no, despite clear evidence they do. *Psychology Today*, June 16, 2020. https://www.psychologytoday.com/us/blog/animal-emotions/202006/do-animals-think-or-feel

Camp, D. (2020). *The honey bee: Understanding the ultimate engineer*. New Holland.

Climate Justice Alliance. (2023). *About Climate Justice Alliance*. https://climatejusticealliance.org/about/

CNN. (2023). *Miliband: "The climate crisis exacerbates poverty."* https://www.cnn.com/videos/world/2023/12/01/exp-david-miliband-cop-intv-fst-12019aseg2-cnni-world.cnn

Derezotes, D. S. (2005). *Spiritually oriented social work practice*. Pearson.

El Akkad, O. (2023). Resiliency in the ashes: As our climate changes, so does the nature of our response. *Orion Magazine*, 18–21, 25. https://orionmagazine.org/article/santiam-fire-2020-detroit-oregon-resiliency/

Encyclopedia of the Environment. (2021). *What is nature?* March 1, 2021. https://www.encyclopedie-environnement.org/en/life/what-is-nature/

Ezell, J. (2023). Climate disruptions are especially dangerous for the opioid epidemic. *Scientific American*, October 3, 2023. https://www.scientificamerican.com/article/climate-disruptions-are-especially-dangerous-for-the-opioid-epidemic/

Fairtrade International. (n.d.). *What is Fairtrade?* https://www.fairtrade.net/about/what-is-fairtrade

Fairtrade International. (2018). *8 ways Fairtrade farmers protect the environment*. June 4, 2018. https://www.fairtrade.org.uk/media-centre/blog/8-ways-fairtrade-farmers-protect-the-environment/

Frankopan, P. (2023). *The Earth transformed: An untold story*. Knopf.

Fritscher, L. (2021). What is senophobia? *Very Well Mind*. https://www.verywellmind.com/xenophobia-fear-of-strangers-2671881

Gelles, D., & Walt, V. (2023). There's a financial paradox blocking efforts to fight climate change. *The New York Times*, December 4, 2023, B2.

Goodell, J. (2023). *The heat will kill you first: life and death on a scorched planet*. Little, Brown, & Company.

GreetingIdeas.com. (2023). *30 best xenophobia quotes phrases to stop discrimination*. https://greetingideas.com/inspirational-xenophobia-quotes-sayings/

Harrabin, R. (2021). *World's wealthiest "at heart of climate problem."* BBC, April 12, 2021. https://www.bbc.com/news/science-environment-56723560

Hinde, D. (2023). Farewell to coal country. *New Humanist*, 22–25.

Jeffers, S. (1997). *Feel the fear ... and do it anyway*. Rider.

Klein, N. (2023). *Doppelganger: A trip into the mirror world*. Farrar, Straus and Giroux.

Krishnamurti, J. (1995). *The book of life*. HarperCollins.

Krishnamurti, J. (1996). *Total freedom: The essential Krishnamurti*. HarperOne.

Kuchta, D. (2023). *Do plants have feelings? A science explainer*. Treehugger, May 4, 2023. https://www.treehugger.com/do-plants-have-feelings-science-explainer-5546944

Mann, T. C., & Ferguson, M. J. (2015). Can we undo our first impressions? The role of reinterpretation in reversing implicit evaluations. *Journal of Personality and Social Psychology, 108*(6), 823–849. https://doi-org.ezproxy.lib.utah.edu/10.1037/pspa0000021

Martinez, D. (2019). Redefining sustainability through kincentric ecology: Reclaiming Indigenous lands, knowledge, and ethics. In M. K. Nelson & D. Shilling (Eds.), *Traditional ecological knowledge: Learning from indigenous practices for environmental sustainability* (p. 140). Cambridge University Press.

McGilchrist, I. (2009). *The master and his emissary: The divided brain and the making of the Western world*. Yale University Press.

Miller, M. C. (2016). The truth about altruism. *Harvard Health Blog*, January 5, 2016. https://www.health.harvard.edu/blog/the-truth-about-altruism-201601058929

Mkwalo, N. (2019). *South African churches take stand against xenophobia*. UM News, October 21, 2019. https://www.umnews.org/en/news/south-africa-churches-take-stand-against-xenophobia

Morgan, R. G. (2020). *Beaver, bison, horse: The traditional knowledge and ecology of the Northern Great*

National Geographic. (2023). Why conquer? https://education.nationalgeographic.org/resource/why-conquer/

Norton, P. J., & Antony, M. M. (2021). *The anti-anxiety program*. Guilford.

Rainey, C. (2023). A whole lot hotter for rich people. Research suggests that the wealthy's luxury shield is rapidly melting. *Fast Company*, December 1, 2023. https://www.fastcompany.com/90990849/climate-change-plant-biodiversity-luxury-effect-rich-poor-inequality

ReligiousTolerance.org. (2021). The golden rule: Ethics of reciprocity. http://www.religioustolerance.org/reciproc2.htm

Roosevelt, E. (n.d.). *Quotations by Eleanor Roosevelt*. Eleanor Roosevelt Papers Project, Columbian College of Arts & Sciences. https://erpapers.columbian.gwu.edu/quotations-eleanor-Roosevelt

Seo, H. (2023). America is getting lonelier and more indoorsy. That's not a coincidence. *The Atlantic*, November 13, 2023. https://www.theatlantic.com/health/archive/2023/11/nature-avoidance-social-isolation-loneliness/675984/

Simmons, D. (2020). What is climate justice? *Yale Climate Connections*, July 29. https://yaleclimateconnections.org/2020/07/what-is-climate-justice/

Sogorea Te' Land Trust. (2023). *What is rematriation?* https://sogoreate-landtrust.org/what-is-rematriation/

St-Pierre, É. (2010). *Fair trade: A human journey*. Quebecor Media Book Group, Inc.

Stirn, M. (2024). When the water dried up. *Archeology Magazine*, January 1, 2024, 38–41.

VanderHoff Boersma, D. (2010). Preface: Fair trade—A constant learning process. In É. St-Pierre, *Fair trade: A human journey* (p. 6). Quebecor Media Book Group, Inc.

Vaughan, F. & Walsh, R. (1988). Gifts from a Course in Miracles. New York: Penguin.

Walsh, S. (2023). *Biden signs order prioritizing "environmental justice."* Associated Press, April 23, 2023. https://www.usnews.com/news/business/articles/2023-04-21/biden-to-sign-order-prioritizing-environmental-justice

Wilkerson, I. (2020). *Caste: The origins of our discontents. New York:* Random House.

Winters, C. (2023). The great guzzlers: Who's using all our water? *YES! Magazine*, 14–15.

Wrong, Y. S. (2023). The forest cleaning: On what forests whisper when we listen. *Orion Magazine*, 8.

Yoder-Grist, K. (2023). This climate crisis map shows how vulnerable your neighborhood is. *Popular Science*, October 7, 2023. https://www.popsci.com/environment/climate-change-threat-map/

Credit

IMG 9.1& 9.2

CHAPTER 9

Climate-Sensitive Practice With Families and Institutions

Nature holds the key to our aesthetic, intellectual, cognitive and even spiritual satisfaction.

—E. O. WILSON (2023)

And into the forest I go, to lose my mind and find my soul.

—JOHN MUIR (2023)

When I was a schoolboy, our parents took us fishing occasionally on a summer Saturday morning. In the already steamy Chicago predawn, they'd wake up my brother and me, and we'd gladly pull on jeans and tie up our gym shoes, our eyes still asleep. The Buick was loaded up before any eastern sky showed early light, and soon we were watching streetlights stream by on old Highway 14. Lake Zurich had an old "seawall" where my brother and I would spend the morning while Dad and Mom picked out a boat for themselves at Bill's for the day.

As we sat on the old seawall, there was just enough light to put a worm on our hooks without impaling our fingers very often, and we cast out our lines into the dim emptiness. As it was still too dark to see a bobber, I had to slowly reel it in to feel a bite.

Then would come a sharp tug, and I'd find myself pulling back on the rod and turning the handle on my reel. I could tell if it was a big bluegill on my line by how the fish fought. Due to its size, it pulled harder than most other fish.

My heart would be beating quickly as I lifted the fish out of the water and swung it up over the old concrete wall and into my hand. It shone in the moonlight as I took the hook out of its mouth and put it in the water bucket.

For a minute I had forgotten about everything else, including pressure to get good grades, dealing with bullies or trouble at home, or the threat of falling nuclear bombs. Nature was the great healer.

But ... I still felt bad for the fish, and for the worms. Even though Dad said they couldn't feel pain, it seems like they maybe did. The worm twisted fiercely as I pierced it. The fish gasped for water through its gills and struggled to get free. Well, we did cook them later.

But there was also this scary underlying feeling that things were not the same as they used to be and that the well-being of the lake ecosystem would continue to worsen over time. We could feel that even at age 10. The old-timers on the pier talked about how good the fishing was 50 years ago.

And even as a young boy, I knew that in 50 years, two children like my brother and me might sit right where we were sitting now, and some old-timer (maybe an older version of me) might say to them, "Back in the 1950s, the water was still so clear that you could see the fish, and you could still safely eat the ones you caught."

Was it, is it still ... inevitable that the quality of our land water and air would continue to decline?

We live in the era of the great climate crisis, when the well-being of all life and our life-supporting ecosystems on our planet are threatened. In this chapter, I first introduce the idea of *climate-sensitive practice.* Then I review climate-sensitive practice with families as well as with institutions.

Climate-Sensitive Practice

Climate-sensitive practice includes the approaches that are described in this text, which are designed to help people not only cope effectively with the stress of the climate crisis but also contribute to the solutions necessary for us to reverse this crisis that threatens all living things on Earth.

I chose the phrase *climate-sensitive practice* to emphasize that the work is informed by our ecobiopsychosocialspiritual *sensitivity* to what is actually happening in our world and in ourselves. Our sensitivity includes not just a cognitive understanding: This is because we live in what some call a post-truth world in which, although our scientists have overwhelmingly demonstrated that the largely human-caused climate crisis exists and can be reversed, there are those who generate disinformation to the contrary for their own political and economic gain. Our sensitivity also includes, therefore, our emotional, physical, social, and spiritual consciousness that tells us every day, if we are paying attention, about what is happening to ourselves, life on Earth, and the ecosystems that support all living things on our planet.

Climate-sensitive practice does *not replace* any traditional practice approaches. However, it *does add* the climate change approaches in our text to the more traditional practice approaches we now currently use. Climate-sensitive practice builds on the principles of inclusive practice that we have studied in the text; therefore, we could say that climate-sensitive practice is a form of inclusive practice that focuses on the climate crisis.

What are the overall goals? Climate-sensitive practice emphasizes the development of people's consciousness of themselves and their world and of their awareness of their connection with other people, other living beings, and the ecosystems that sustains all life. We also encourage people to participate in dialogue with other people so that they can bridge the differences that currently divide us.

What are the main practice principles of climate-sensitive practice? Regardless of the work setting (e.g., mental health, physical health, education, corrections) and the professional role (e.g., psychotherapist, nurse, teacher, corrections officer), climate-sensitive practice includes the principles listed in Table 9.1.

TABLE 9.1 Principles of Climate-Sensitive Practice

Practice Phases	Practice Principles
Engagement	1. Identify, normalize, and support how people experience the climate crisis
Assessment and evaluation	2. Identify and acknowledge the person's hopes and fears about the climate crisis 3. Assess the current climate of the individual, group, or community at the ecobiopsychosocialspiritual levels of development 4. Identify and address climate justice issues 5. Inform climate-sensitive practice with both practice-based evidence and evidence-based practice methods
Intervention	6. Work on climate change using all ways of knowing, including science, logic, intuition, and imagination 7. Foster climate change by assisting in the healing and transformation of the individual and collective ego 8. Encourage climate change through consciousness work, dialogue practice, and faith-based activism 9. Foster climate change at the micro, mezzo, macro, and eco levels 10. Use all seven paradigms of inclusive practice in climate change interventions

Practice Elements and Principles

I find that it is usually not helpful to think about the four practice phases as distinct elements that always follow in a linear order. Rather, it seems more helpful to think about them as interconnected parts of a whole that do not necessary need to happen in a particular order. Therefore, the phases of engagement, assessment, intervention, and evaluation can all occur over and over again and in different order and can all show up briefly even in the same session from time to time. I combine assessment and evaluation in Table 9.1 because I find I use similar techniques in conducting both. What follows next are brief descriptions of each of the 10 principles listed in Table 9.1.

Engagement

1. *Identify, normalize, and support how people experience the climate crisis*

Many people are not yet aware of their deeper emotions, feelings, and beliefs related to the climate crisis. When asked, they may first say things like, "Hey, that's all a hoax" or "Oh, I try not to think about it." It may be helpful to ask people specific questions, such as "What basic emotions do you have, including sad, mad, glad, scared, excited, or excited, about the climate crisis?" Another question might be "Whom do you tend to blame for what is happening?" which is more of a question about beliefs. We usually want to let them know that their reactions are both normal for them and also probably common and shared by many others.

Sometimes we might need to ask them directly about their experience of the climate crisis, because they may tend to deny or minimize that experience. I might ask people how much they believe that the climate crisis currently affects their physical and mental well-being. One of my clients told me that they think climate news contributes 50% to their daily depression and anxiety. I want to understand whether they feel empowered or perhaps helpless and in despair.

We want to offer support for people when they first start talking about these feelings and beliefs. When they share them, we do not have to agree with what they say to offer support, but we can respond by saying something like, “I don’t blame you for feeling angry and helpless” or “I can understand why you want to blame the politicians.”

Assessment and Evaluation

2. *Identify and acknowledge the person’s hopes and fears about the climate crisis*

Often it is helpful to ask people about their hopes and fears about climate. Many of us seldom, if ever, have an opportunity to discuss these hopes and fears in a safe environment in which we can truly explore and share them. I find that when I hear about these hopes and fears, it usually also helps inform me more about what interventions to use with them.

We might also want to ask more questions about what people are most worried about. Many of us probably want to determine how the changing climate will affect our immediate lives; for example, is my home at risk of fire or flooding, and will safe and clean food, water, and air be available to me and my family? These most pressing and immediate concerns will probably need to be addressed early in our work with our students, clients, patients, and other community members.

We also want to explore what people hope for. Do they want, for example, to find ways to make a positive difference in their world? Or do they actually want to continue minimizing or denying the climate crisis we all face? We help them identify their hopes and begin talking about how they might begin achieving them.

3. *Assess the current climate of the individual, group, or community at the ecobiopsychosocialspiritual levels of development*

I keep a mental list in my head of all the dimensions of human development and try to eventually ask my clients, students, or patients at least one question related to each of them. For example, on the emotional level, I might ask them how they feel about the loss of species diversity in their local community, and, at the cognitive level, I might ask what they think about the loss of biological diversity in their local community.

I also want to assess which of the dimensions of development each client, patient, or student is most advanced in and which ones are the least developed. For example, the person may be very skilled in describing their reactions to the climate crisis intellectually and not as comfortable describing their emotional reactions.

These assessments help inform my interventions; for example, I might want to start by working with the developmental dimension with which they are most confident and comfortable. It is often effective to first build on the person’s strengths. For example, most people are probably comfortable initially discussing things at a cognitive level, so I often begin with an intellectual conversation about what is going on.

4. *Identify and address climate justice issues*

I want to help the community members with whom I work identify the climate justice issues in their life. I want them to explore the ways in which they may be privileged as well as how they may be minoritized in the world. I also want to understand how much support they have in their world, both formal (e.g., counselors, teachers, nurses) and informal (e.g., neighbors, friends, families) as well as the financial resources available to them.

Some clients may need to hear what climate justice is about before they can talk about it. Sometimes I like to refer people to current and readable materials that they can review. For example, I recently recommended a wonderful interview of the "father of environmental justice" by Robert D. Bullard in *Scientific American* (Funes, 2023) to a number of students, community members, and students. The article focuses on how racial inequality created a "geography of injustice" in the United States.

5. *Inform climate-sensitive practice with both practice-based evidence and evidence-based practice methods*

In terms of practice-based evidence, I find that it is helpful and important to ask people for feedback during most meetings I have with them. The request can be quite simple, such as "Did you find today's meeting helpful? Please explain." It is also important to show them how I intend to follow-up and respond to their feedback. In terms of evidence-based practice, I like to read the literature and strive to be an up-to-date and critical consumer of what I read, looking for new theories and new data that might help inform my practice. In addition, I want to use the best science-based information I can find when I make presentations about the climate crisis in my work.

Intervention

6. *Work on climate change using all ways of knowing, including science, logic, intuition, and imagination*

As I make decisions about how to intervene with an individual or group, I strive to make choices based on all the ways of knowing available to me. The more that my science, logic, intuition, and imagination agree, the more confident I am in my assessment. For example, when a person talks about their issues, I might initially have a strong countertransference feeling of impatience with them. I know that this feeling means something; therefore, I might additionally use my logic to ask myself various questions, such as "Is it possible that this person is also actually impatient with themselves and others?" I might also have read an article about impatience recently, which can help inform my responses. Finally, I can use my imagination to consider the various interventions I might try with the person in relationship with the theme of impatience.

7. *Foster climate change by assisting in the healing and transformation of the individual and collective ego*

I often introduce the concept of ego and such related concepts as shame, grandiosity, and left-hemispheric thinking early in my work with people so that they can start to develop an understanding of what these terms mean and hopefully a greater comfort with talking about them. And as the person becomes increasingly ready and able to talk about their ego, for example, I want to be supportive and normalize their experience and generally encourage their increased awareness and acceptance of the functions of their ego processes.

8. *Encourage climate change through consciousness work, dialogue practice, and faith-based activism*

In general, I want to support my patients, students, and clients in learning about the nature and benefits of consciousness work, dialogue practice, and faith-based activism. Often I find that the best way to teach about these methods is experiential, and I might strive to find ways to engage them in these activities so that they can learn to participate in them experientially.

9. *Foster climate change at the micro, mezzo, macro, and eco levels*

We have studied these levels in our text. As professional helpers, we have our own preferences about how we want to do our own climate activism; some of us may prefer macro-level work, whereas others favor the micro level. We know that when we are making a difference in the world, we are modeling positive behaviors for our clients, students, and patients and other community members.

We all have these kinds of preferences, of course. For example, some of the people with whom I work in my community are very open to learning about how they can help make a climate change in their own consciousness, whereas others are less so. Some are more eager to participate in fostering climate change in their families, institutions, and communities, whereas others may prefer working with local humansystems or ecosystems. I encourage everyone to think about where and how they want to become a faith-based activist, and I remind them that everyone can make a difference somehow in the world.

10. *Use all seven paradigms of inclusive practice in climate change interventions*

In the text, I have also provided information on the seven paradigms of practice (e.g., see Table 9.4). I keep a mental list in my head of these paradigms, which helps me as I think about how to intervene. I often use several interventions, each drawn from different paradigms, in the same session, based on my evolving assessment of the person and their world. As I experiment with these interventions, I continue to learn more about which ones are most effective in which situations.

Climate-Sensitive Practice With Families

We helping professionals often work with families, sometimes seeing family members together or individually, as a couple, or even with other families in larger-sized groups. We can think of the families we see as microcosms of the human family. In most families, just like in our local and global communities, there are usually struggles for faith, love, power, fame, and wealth. In addition, the climate crisis has affected most families in a variety of ways, including regarding our ecobiopsychosocialspiritual well-being, economic changes, political policies, and polarization within and around the family. And many family members feel discouraged and even despair about the future and sadly become disempowered to do anything to change the world.

When my social work students worry about whether they are doing enough to reverse the impact of the climate crisis, I remind them that every time they work with an individual or family, they can make a difference in the world. When we help people act more civilly with each other, listen to each other, and treat each other with loving kindness, we may find that they also become more attuned to all living things and our ecosystems. When I become conscious of and start to temper my ego, I may also begin to behave differently in all realms of my life. Every time, for example, we help end domestic violence or child maltreatment in a family, we probably help improve the lives of many people, who then may go on to live more responsible and loving lives in their community

In this section, we will look at climate-sensitive approaches to work with families, using the assessment and intervention strategies of inclusive practice that we introduced in Chapter 1. Tables 9.2 and 9.3 build on Tables 1.1 and 1.2 from earlier in the text. In assessing a family who is living in our current climate crisis, helping professionals can use Table 9.2 to help guide them in how to assess a family's ecobiopsychosocialspiritual climate.

TABLE 9.2 Ecobiopsychosocialspiritual Family Assessment in Our Work With Humansystem Climate Change

Dimension	Basic Focus	Examples of Assessment Questions in Climate-Sensitive Practice
Eco (ecological)	Our relationship with our human-made and "natural" environments	How does the family's local culture and community view the climate crisis? What are the local impacts of the climate crisis on the air, water, and soil?
Bio (biological)	Our relationship with our body	What attitudes does the family have toward their bodies? How has the climate crisis affected the physical health of family members?
Psycho (psychological)	Our relationship with our emotions and thoughts	What early experiences and feelings did they have in nature? What do family members feel about the climate crisis? Do family members feel despair or exhibit faith-based activism about the climate crisis?
Social	Our relationship with other people	What kind of relationships, if any, do family members have with people who disagree with them?
Spiritual	Our relationship with our own spirituality	What kind of connection do family members have with other living things and ecosystems?

For example, let's say that Samantha South, a licensed clinical social worker, is working with the Smyth family. Mrs. and Mr. Smyth have two children, Suzie, who age is 13, and Sammy, who is age 10. Let's suppose that the presenting problem is that little Sammy is showing signs of depression and anxiety and is refusing to go to school over at Puppy Hill Elementary. How can Samantha include climate change issues in her assessment?

It turns out that Mrs. Sherri Smyth is age 44 and a successful accountant in her own private business downtown, about 30 miles from their suburban home. Mr. Sid Smyth is age 42 and does local yard work in the summers and snow removal in the winters. Partly since his schedule is less demanding, he is the parent who does most of the housework and everyday care of the children. The family lives in a mountain valley suburb of a midsized city in the Great Basin. When asked, they report that they have lived on the Wasatch Front all of their lives and that everything seems more crowded and intense now, including freeways, lines to get into local ski resorts, parking lots for trailheads, and drop-off lines for the local Puppy Hill Middle School.

Samantha asks Sherri and Sid to meet with her first to discuss the family. They decide to have a teletherapy meeting, using videoconferencing technology. Samantha discovers that Sid is very frustrated with and worried about little Sammy. Sid explains that Sammy starts crying every morning and wants to stay home, and he complains that he has "tried everything to get Sammy to go to school." Samantha assesses that Sid and Sammy have a more "enmeshed" (very close) relationship in the family system and wonders whether Sammy may actually be rewarded for not going to school by his father's worried and sometimes overprotective reactions. In contrast, Samantha also assesses that Sherri may be more "disengaged" (very distant) with Little Sammy and rarely offers him any nurturing. When asked, both parents say that Suzie is doing fine, is getting straight A's, and is the star violinist in the school orchestra.

The couple denies having any marital issues, but Sherri reports, "I am so busy, and with the long commute, I'm exhausted by the time I get home in the evening and hardly see Sid anymore." Samantha's first impression is that this couple is disengaged with each other.

Samantha later sees Little Sammy in a face-to-face session in her office. The boy presents as shy, unable to sit still in his chair, and slightly below average in weight and height for his age. He reports that he "hates school" but really likes playing electronic games with his father.

Since Samantha cares about the climate crisis and has read our text, she believes in using climate-sensitive practice. She knows the family cares about the environment and has heard them talk about how much they enjoy hiking in the summer and skiing in the winter and how much they hate the December and January valley atmospheric inversions. She is not certain about the extent to which environmental factors are associated with Little Sammy's depression and anxiety, but she does think that such aspects as poor air quality, long commutes, global warming, and summer heat waves affect the entire family. A summary of Samantha's assessment of climate crisis–related issues is shown in Table 9.3.

TABLE 9.3 Samantha's Assessment of Climate Crisis–Related Issues in Smyth Family Example

Dimension	Assessment Issues Identified in Smyth Family Example
Eco (ecological)	Family in ecological stress from deterioration of air, water, and earth quality Parents have to face long commute into town Family exposed to constant news about climate crisis Local community having more frequent and severe weather extremes, including heat waves, droughts, and flooding
Bio (biological)	Family physical well-being affected by air, water, and earth pollution
Psycho (psychological)	Some of the family have symptoms of depression and anxiety associated with constant climate crisis news and local conditions
Social	Political polarization in family and local and global communities affect family members
Spiritual	A general sense that the quality of life is gradually worsening because of climate crisis–related issues and a related sense of despair are associated with a gradual loss of connection with nature

Samantha decides to offer a number of interventions to this family, some of which are based on traditional family therapy strategies. For example, she encourages Sid to stop rewarding little Sammy with attention when he cries in the morning and encourages Sheri to spend more time with Little Sammy. Samantha also asks the parents to focus a little more on Suzie's mental health. Samantha also helps the parents find ways to nurture their marriage, including by regularly scheduling dates on weekends, such as for long walks or hikes outdoors.

Some of the interventions are climate-sensitive, and these are broadly summarized in Table 9.4. When possible, Samantha wisely combines climate-sensitive with traditional interventions.

TABLE 9.4 Paradigms of Social Work Practice

Paradigm	Basic Goals	Target Dimensions	Examples of Climate-Sensitive Interventions
Psychodynamic	To address past trauma so that I can better meet my needs now	Cognitive and social	Explore impact of increasing pollution and crowding over the years
Cognitive-behavioral	To change the way I think and act in the here and now	Cognitive and social	Challenge the despair the family feels and encourage faith-based activism
Experiential-phenomenological	To increase awareness and effective expression of emotions	Emotional	Explore feelings that family members have about the environmental quality of their life
Transpersonal	To experience my connection with myself, other people, other living things, and ecosystems	Spiritual	Explore the connection that family members have with each other, with all living things, and with the ecosystems that support us all
Case management	To receive advocacy, information, and referrals	All	Refer family members to local resources, including community gardens, car pools, hiking clubs
Biopsychosocial	To gain awareness, acceptance, and healing of physical body	Biological	Encourage outdoor aerobic exercise for family members, such as Sherri going on summer hikes or winter ski dates with her son
Ecobiopsychosocial (or deep ecology)	To take response-ability for the well-being of other people, other living things, and ecosystems	Ecological	Encourage involvement in community gardening and faith-based climate activism

Biopsychosocial Education

Another way to work with families is through biopsychosocial education. In our text, we can define *biopsychosocial education* as the provision of knowledge and resources to clients, students, and patients that can help inform and support their resiliency, biopsychosocial well-being, and faith-based activism in responding to the climate crisis. In this section, we will explore some strategies that helping professionals can use with the families with whom we work. Not all clients will be open to such education, of course; I have learned to back off when the client becomes resistant or reactive and usually stop teaching and ask them to discuss what they are feeling and why.

For example, we may need to pay better attention to the language we use to talk with people about the climate crisis. Cale Kennedy (2023), the chief technology officer of Carbon Direct, suggests that we use the phrase *climate opportunity* when talking about the climate crisis. He writes:

> There are many ways to "treat" climate change. The bad news is that for people in the communities most vulnerable to its impacts, the treatment options offered by experts—eat less meat, use less electricity, buy a smaller car, stop drilling for oil—are often impractical. Talking about climate change as a condition with many treatment options, including many that have positive outcomes and co-benefits for these communities that extend beyond climate, can be effective in reframing the conversation. ... Communicate opportunities, not sacrifices. (pp. 1–2)

We can be more mindful of how we talk with people about the climate crisis and strive to be more positive when possible.

In the story about the Smyth family, we mentioned that the mother has a long commute to get to work. Helping professionals can help people understand that long commutes can have multiple negative impacts on biopsychosocial health. In addition to requiring time, long daily commutes are associated with increased weight gain, alcohol consumption, lack of exercise, and sleep problems. Researchers have also found links between long commutes and depression (Watson, 2023). There are, in fact, some wonderful alternatives to car commuting, including opportunities to bicycle, walk, take a bus or commuter train, and carpool. The aerobic exercise involved in daily bicycle commuting, for example, can bring many health benefits.

We can also educate people about the negative impact of many common household activities and products upon our well-being. For example, regular exposure to the use of wood burning to heat homes has been found to be associated with significant health problems:

> A recent study estimated that 284 Londoners a year are dying early due to outdoor air pollution from solid fuel heating. It also estimated that about 90 new cases of asthma in children, 60 new cases of stroke and 30 new lung cancer cases a year were linked to this pollution. This is a health burden of almost £800 a year created by the average person in London who uses a fireplace or stove. (Fuller, 2023)

We can help people in the community identify the potential risks in engaging in various activities or using various products and then enjoy the long-term health outcomes that can come from making different and better decisions about their homes and activities.

Our work in biopsychosocial education is especially important in an era when misinformation or censorship is employed by businesses and governmental agencies. In fact, the presence of dangerous forever chemicals in our environment and the lack of regulation of dangerous chemicals in the United States were named as two of the top five censored news stories of 2023 (Rosenberg, 2023). Forever chemicals take a long time to break down, which allows them to increase inside of humans, other living things, and the environment. Perfluoroalkyl and polyfluoroalkyl substances (PFAS), for example, are forever chemicals found in our rainwater, surface water, and soil in many areas in which humans live. PFAS have been linked to various kinds of human cancers, developmental delays in children, fertility issues in adults, reduced vaccine efficacy, and high cholesterol levels. And toxic chemicals, including forever chemicals, go mostly unregulated in the United States because the chemicals industry had a hand in writing the 1976 Toxic Substances Control Act (TSCA) and because many industry-friendly scientists work for the Environmental Protection Agency as regulators (Rosenberg, 2023).

Another example of emerging science that can help our families is the study of where weather extremes such as floods and fires are most likely to happen. Many residents or future residents may not know that they are living nor wish to live in a high-risk zone (*The Economist*, 2023). When families are knowledgeable of such risks, they can make decisions that will benefit them for the rest of their lives.

Climate-Sensitive Practice With Institutions

Helping professionals may often have opportunities to work directly or indirectly with local institutions, including businesses, churches, and schools. We can assist institutions in making the climate changes necessary to help protect our local communities from the climate crisis and support faith-based climate activism. We will often use the same phases of engagement, assessment, intervention, and evaluation in working with institutions, but the work we do may be located in our offices, at the institution itself, and/or in the larger community. In this section, we will look at examples of work in several different locations.

Businesses

Researchers have found that most people in the United States want businesses and corporations to do more to address climate crisis impacts. Two-thirds of adults say large businesses and corporations are not doing enough to reduce these effects. Only 21% say they are doing enough, and only 10% say they are doing too much (Tyson et al., 2023).

The information provided earlier in this chapter on forever chemicals and the lack of regulation of toxic chemicals by the United States (Rosenberg, 2023) can, for example, help inform community members about what products they should buy, and which they should not. Helping professionals can support consumers empower themselves into educated consumers who buy products from businesses and corporations who actually strive to protect people, other living things, and the environment from dangerous chemicals.

We can also assist local organizations revise their mission statements. As Damon and Bell (2023) recently stated, "The ever-increasing threats of climate change challenge the mandate of every organization" (2023, p. 59). Helping professionals can offer local seminars and dialogues for community members where such topics can be discussed openly. We can also refer leaders to sources of additional information, such as the Stanford Doerr School of Sustainability (2023), where interested participants can learn more about sustainability efforts and mandates.

In my work with businesses and corporations, I have found that most people in these organizations want them to become more sensitive to environmental concerns. They are often frustrated that the efforts they have made to protect the environment are neither appreciated by consumers nor by their employers. However, there are some business leaders who always seem to put profit over sustainability and who unfortunately attempt to "greenwash" their company and products.

According to earth.org (Robinson, 2023):

> Greenwashing is essentially when a company or organization spends more time and money on marketing themselves as being sustainable than on actually minimizing their environmental impact. It's a deceitful advertising method to gain favor with consumers who choose to support businesses that care about bettering the planet. Greenwashing takes up valuable space in the fight against environmental issues, like climate change, plastic ocean pollution, air pollution and global species extinctions. (p. 1)

Business leaders know that if they are seen by consumers as ethical, they can make more money. Table 9.4 provides a list of marketing approaches that may be associated with greenwashing, which we can provide to consumers.

TABLE 9.5 Marketing Approaches That May Be Associated With Greenwashing

1. The use of "fluffy" language with no clear meaning, such as *eco-friendly*
2. Statements that a company is slightly greener than its competitors
3. "Greening" dangerous products to make them seem safe, such as *eco-friendly* cigarettes
4. Using language that only a scientist could review or understand
5. Providing no proof of green claims
6. Presenting lies as facts
7. Showing one small green feature when almost nothing else is green
8. Generally refusing to be transparent or open or to admit mistakes

Source: Robinson, 2023.

In *This Changes Everything: Capitalism vs. the Climate*, Naomi Klein (2015) linked our capitalistic economic system, which is based on fossil fuels, with both the climate crisis and the difficulties in our social and political systems. In Chapter 10, we will examine this perspective at a macro level in greater detail.

In our work with local communities, Klein's work can help us understand the socioeconomic and political reasons why the business world and larger culture may resist the kinds of climate change work that we have explored in the text. Klein points out how a central conflict exists between the needs of our present economy and the needs of the life on Earth: Whereas our economic model requires unregulated growth of resource extraction, our climate crisis requires a reduction in our use of resources.

Schools

The vast majority of us went to school when we were children and adolescents. Education is a key foundation of a healthy community, and our schools are important educational institutions in our communities. We know that educational background is a significant factor associated with how people think and act regarding the climate crisis. Those who are more educated, female, and/or younger are more likely to see climate crisis as a serious threat (Fagan & Huang, 2019).

As a former school teacher and professor, I continue to have strong feelings about the importance of education in our world. I think that affordable (or, even better, free) schools should be open to everyone throughout their lifetime and that vital issues of the day should be discussed in our classrooms. Hopefully, we can all support the teaching of science-based knowledge in our schools as well as the teaching of consciousness work and dialogue techniques to both children and adolescents.

Over the years, for example, I have participated in efforts to create dialogue "teams" in local high schools. Although most U.S. high schools have debate teams, very few, if any, have dialogue teams, despite the fact that dialogue has so much potential as a form of communication that could help heal and transform our local and global cultures. I like the model of teaching dialogue facilitation skills to high school students and then having them co-facilitate conversations about important social issues with other students in their school and other community schools (including at middle and grade schools). High school students often seem to grasp the dialogue philosophy quickly and understand the need to expand dialogues in our schools and communities.

Students and their teachers have had had to deal with increasing numbers of and increasingly severe climate crisis–related challenges in the past few years:

> An increase of natural disasters from wildfires to floods to hurricanes to tornadoes—exacerbated by climate change—have ravaged America's schools since students returned to in-person learning after the COVID-19 pandemic. And manmade disasters from lead in drinking water to asbestos in school buildings are playing a role. (Jimenez, 2023, p. 1)

Researchers have found that school closures extending to 10 days were "generally insignificant" except for the harmful impact of wildfire closures on grade school students. Older students seemed more resilient, however, to the closures, as demonstrated by their score performance. Many students experienced additional stress after severe weather events when their families moved away or encountered inequities in recovery funding (Jimenez, 2023). Helping professionals can provide support and resources for both students and teachers as well as parents who have suffered from these kinds of stressors.

Unfortunately, schools and educational materials have also been under another kind of assault. Requests to ban books in school and public libraries are now at a 21-year high, with topics that challenge gender stereotypes and racism among the most banned (Leipzig, 2023). Such requests are often associated with the polarization that we discussed in Chapter 7, and library materials about the climate crisis are also vulnerable to such requests. These demands tend to be most prevalent in the most conservative counties and states. Helping professionals can host public dialogues in which all citizens, including those conservative, liberal, and in between, gather to talk civilly about our hopes and fears for our children. I have found that such dialogues can lead to greater mutual understanding and trust between politically polarized groups. As understanding and trust develop, perhaps we can avoid taking the more drastic step of removing from our libraries books we do not like.

Churches

Many of us belong to religious organizations or work with people who belong to one, and most of us have opportunities to help these institutions heal and change from within. Members of religious organizations can bring up climate issues with other members and speak up for faith-based climate activism. We have examined in earlier chapters how religiosity is associated with views about the climate crisis. Many religious organizations have subgroups of members who disagree about what they think and feel about the climate crisis.

For example, a recent Reuters article (Valdmanis, 2023) describes how Pope Francis advanced his new environmental stewardship program in 2015 and asked for the reduction of fossil fuel burning. However, the article reveals that in the United States:

> not a single diocese has announced it has let go of its fossil fuel assets. U.S. dioceses hold millions of dollars of stock in fossil fuel companies through portfolios intended to fund church operations and pay clergy salaries, according to a recent Reuters review of financial statements. And at least a dozen are also leasing land to drillers, according to land records. (p. 1)

As a young churchgoer in my childhood, I observed the powerful impact religion can have on all of us and the support that a religious community can provide for people who are in suffering. My hope is that our religious communities can work together with the larger communities to address our climate crisis.

A recent article by CBS News described how a group of young Evangelicals challenged the prevailing attitudes in their church. Currently, about 53% of Americans say that human activity is responsible for a warming planet, but only 32% of Evangelical Christians agree with them. The Evangelical church did not always oppose climate crisis work in the United States. Evangelical Christians in the 1970s were early leaders in raising concerns about environmental degradation, but in the 1990s, political conservatives began to put economic growth over environmental concerns by casting doubt on climate science. In 2022, Galen Carey, vice president of government relations at the National Association of Evangelicals, began using a biblical approach to encourage church members to help check or mitigate climate change (Breen et al., 2023).

The young woman highlighted in the CBS article, Elsa Barron, is a climate research fellow who works to educate peers in the Evangelical Christian church about the climate crisis. She said:

> What does loving our neighbors really look like in a world where the sorts of decisions are directly impacting people's ability to live in their homes across the world, or to manage their crops or have food or water to drink? Well, very sadly, this whole issue has become politicized in an unhelpful way. And they say, "Oh, well, if I am conservative then, or if I'm Republican, or whatever, then I must be opposed to this stuff." I think that's, unfortunately, where the issue [has] wound up for a lot of people. (Breen et al., 2023, pp. 1–2)

Breen also states that she wants to continue to open spaces in which conversation can happen and meet the family members and church members who may resist her efforts with compassion and kindness. This story is heartening because it suggests that we can find ways to bridge the cultural or religious differences that divide us and cooperate to become better caretakers of our planet.

As we discussed in earlier chapters, there is evidence that the majority of people support such sustainability goals as the reduction of greenhouse gas emissions and the protection of wild animals and ecosystems. Helping professionals can help educate our communities about this fact, which can help encourage faith-based activism in our communities. In fact, Convention on Biological Diversity Acting Executive Secretary David Cooper says that it is politicians, rather than the public, who fight against the "green agenda" because their parties are "seeking 'wedge issues' for electoral gain" (Greenfield, 2023, p. 1). For example, a majority (two-thirds) of Americans support making the development of renewable energy a priority. About 30% of the United States' population want to phase out fossil fuels altogether, with younger adults most in support. And approximately 68% want to see a mix of energy sources that includes fossil fuels. Researchers have found that a majority of people worldwide are also concerned about the climate crisis, although few countries are as politically polarized about the topic as the United States (Tyson et al., 2023).

Supporting Democracy at an Institutional Level

Professional helpers know that it is human nature to want to minimize or deny difficult emotions or information, and we also recognize that people and organizations in democracies have a right to have their own opinions. We also know, however, that there will always be leaders who are willing to take advantage of this human tendency to deny bad news and publicly argue against the existence of our climate crisis for their own political and personal gain.

Democracies are vulnerable to those who find ways to take advantage of human fear in ways that serve their own immediate political gain, regardless of the lack of evidence for their stance.

> Conspiracies are thriving online, according to a report by the coalition Climate Action Against Disinformation released last month, in time for the [United Nations] climate conference in Dubai. Over the past year, posts with the hashtag #climatescam have gotten more likes and retweets on the platform known as X than ones with #climatecrisis or #climateemergency. ... People buy into bad information for different reasons, said Andy Norman, an author and philosopher who co-founded the Mental Immunity Project, which aims to protect people from manipulative information. Due to quirks of psychology, people can end up overlooking inconvenient facts when confronted with arguments that support their beliefs. (Toder, 2023)

For example, recent efforts by the Biden administration to regulate dangerous emissions from our air conditioners and refrigerators have been opposed by some, despite the fact that scientists have found that such emissions are dangerous to the well-being of life on Earth (Catenacci, 2023).

Helping professionals can help community members get together and engage in civil dialogue about important issues like climate crisis. We can also, as Yascha Mounk (2022) suggested, discuss how in a healthy democracy people have a responsibility to adhere to higher values in which we put the overall well-being of society ahead of politics or the short-term economic gain of their person or organization. It is the responsibility of all of us fortunate to live in a democracy to remain vigilant and participate in the governance of our families, institutions, and communities.

QUESTIONS FOR REFLECTION

1. What kinds of stories do you have about how your family of origin thought about other living things and the environment? For example, was there a reverence for life and nature?

2. If you are a therapist or psychotherapist, have you had opportunities to include content on the climate crisis into your assessments and interventions? What did you try? How did it work?
3. If you are a teacher, what kinds of conversations have you had about the climate crisis in your classrooms? What kinds of teaching strategies have you used? How did it work out?
4. If you work in health sciences, have you had opportunities to discuss the climate crisis with your patients? Have you found ways to help patients find associations between their well-being and that of the environment around them?
5. After reading about work being done in schools, churches, and businesses, can you identify an institution in your community with which you might like to work? Please describe an assessment that details the climate crisis issues that exist in this institution and then share what kinds of interventions you might try to implement to address those issues.

REFERENCES

Breen, K., Yamaguchi, A., & Mirman, G. (2023). *Young Evangelicals fight climate change from inside the church: 'We can solve this crisis in multiple ways.'* CBS News, September 30, 2023. https://www.cbsnews.com/news/evangelicals-climate-change-fight-inside-the-church/?ftag=CNM-00-10aac3a

Catenacci, T. (2023). *Biden admin issues eco regulations impacting air conditioners, refrigerators.* Fox News, October 11, 2023. https://www.foxnews.com/politics/biden-admin-issues-eco-regulations-impacting-air-conditioners-refrigerators

Damon, M., & Bell, A. (2023). Find your climate mission. *Stanford Social Innovation Review*, 59–60.

The Economist. (2023). Climate change: Uninsurable America: Parts of America are becoming insurable. September 23, 2023, 23–24.

Fagan, M., & Huang, C. (2019). *A look at how people around the world view climate change.* Pew Research Center, April 18, 2019. https://www.pewresearch.org/short-reads/2019/04/18/a-look-at-how-people-around-the-world-view-climate-change/

Fuller, G. (2023). The health cost of burning wood to warm homes. *The Guardian*, December 15, 2023. https://www.theguardian.com/environment/2023/dec/15/pollutionwatch-health-cost-of-burning-wood-to-warm-homes?CMP=oth_b-aplnews_d-1

Funes, Y. (2023). The geography of injustice. *Scientific American*, S3–S7.

Goodreads. (2023). *This changes everything quotes.* https://www.goodreads.com/work/quotes/41247321-this-changes-everything-capitalism-vs-the-climate

Greenfield, P. (2023). Politicians, not public, drive U-turns on green agenda, says UN biodiversity chief. *The Guardian*, October 11, 2023. https://www.theguardian.com/environment/2023/oct/11/politicians-not-public-driving-green-uturns-says-un-biodiversity-chief-david-cooper?CMP=oth_b-aplnews_d-1

Hillman, J. (2023). *We've had a hundred years of psychotherapy and the world's getting worse.* https://www.goodreads.com/author/quotes/18521.James_Hillman

Jimenez, K. (2023). Deadly disasters are ravaging school communities in growing numbers. Is there hope ahead? *USA Today*, September 24, 2023. https://www.usatoday.com/story/news/nation/2023/09/24/natural-manmade-disasters-close-schools/70767671007

Kennedy, C. (2023). Stop talking about the climate crisis and start talking about the climate opportunity. *Fast Company*, December 15, 2023. https://www.fastcompany.com/90998687/stop-talking-about-the-climate-crisis-and-start-talking-about-the-climate-opportunity

Klein, N. (2015). *This changes everything: Capitalism vs. the climate.* Simon & Schuster.

Leipzig, P. (2023). *Requests to ban books hit a 21-year high. See which titles were the most challenged.* CNN, August 27, 2023. https://www.cnn.com/2023/08/27/us/school-library- book-ban-increase-dg/index.html

Mounk, Y. (2022). *The great experiment: Why diverse democracies fall apart and how they can endure.* Penguin Random House.

Muir, J. (2023). *John Muir quotes.* https://www.adventurehiketravel.com/25-john-muir-quotes- inspire-wanderlust/

Robinson, D. (2023). *Explainer: What is greenwashing and how to avoid it.* Earth.org, November 13, 2023. https://earth.org/what-is-greenwashing/

Rosenberg, P. (2023). Project censored. *City Weekly*, December 21, 2023, 12–13.

Stanford Doerr School of Sustainability. (2023). https://online.stanford.edu/schools- centers/stanford-doerr-school-sustainability

Toder, K. (2023). Why fake news about climate change is still so effective: 'It's a lot easier and cheaper to push doubt than to push certainty.' *Mother Jones*, December 24, 2023. https://www.motherjones.com/environment/2023/12/why-fake-news-about-climate-change-is-still-so-effective/

Tyson, A., Funk, C., & Kennedy, B. (2023). *What the data says about Americans' views of climate change.* Pew Research Center, August 9, 2023. https://www.pewresearch.org/short-reads/2023/08/09/what-the-data-says-about-americans-views-of-climate-change/

Valdmanis, R. (2023). *Insight: Defying Pope's calls for climate action, U.S. Catholic bishops cling to fossil fuels.* Reuters, November 29, 2023. https://www.reuters.com/business/environment/defying-popes-calls-climate-action-us-catholic-bishops-cling-fossil-fuels-2023-11-29/

Watson, C. (2023). Massive study finds a link between commuting and poor mental health. *Science Alert*, December 11, 2023. https://www.sciencealert.com/massive-study-finds-a-link-between-commuting-and-poor-mental-health

Wilson, E. O. (2023). *Healing power of nature quotes.* https://www.outofstress.com/healing- power-of-nature-quotes/

Credits

IMG 10.1

Climate-Sensitive Practice With Local and Global Communities

I began to see all kinds of ways that climate change could become a catalyzing force for positive change—how it could how it could be the best argument that progressives ever had to demand the rebuilding and reviving of local economies, to reclaim our democracies from corrosive corporate influence, to block harmful new free trade deals and rewrite old ones, to invest in starving public infrastructure like energy and water, to remake our sick agricultural system into something much healthier, to open our borders to migrants whose displacement is linked to climate impacts, to finally respect Indigenous land rights—all of which would help to end grotesque levels of inequality within our nations and between them … [but] climate change can [also] be a catalyst for a range of very different and far less desirable forms of social, political, and economic transformation. … The crisis will once again be seized upon to hand over yet more resources to the 1 percent … [through] privatization of the public sphere, deregulation of the corporate sector, and lower corporate taxation, paid for with cuts to public spending. … The core problem was that the stranglehold that market logic secured over public life in this period made the most direct and obvious climate responses seem politically heretical.

—NAOMI KLEIN (2014, PP. 7–8, 19)

The health of our societies depends on sustaining a delicate balance between the economic and the political, the individual and the collective, the national and the global. But that balance is broken. Our economy has destabilized our politics and [vice] versa. We are no longer able to combine the operations of the market economy with stable liberal democracy. A big part of the reason for this is that the economy is not delivering the security and widely shared prosperity expected by large parts of our societies. One symptom of this disappointment is a widespread loss of confidence in elites. Another is rising populism and authoritarianism. Another is the rise of identity politics of both left and right. Yet another is loss of trust in the notion of truth. Once this last happens, the possibility of informed and rational debate among citizens, the very foundation of society, has evaporated. … Democracy is always imperfect. But tyranny is never the answer.

—MARTIN WOLF (2023, PP. XIX, XXI)

It's spring 1977, and my Volkswagen Rabbit takes me over parched plains, still-snowy cordilleras, and sage deserts toward San Francisco.

My route, I-80, is nothing new but rather a greatly expanded and paved-over version of the ancient migration routes of buffalo and antelope, then a pathway for Indigenous people and, much later, a primitive migration route for early European immigrants.

These are different times. Most of what we now call North America has been transformed from ancient biomes to what academics call *anthromes*, or human-altered landscapes. The wilderness now is in our minds, a mental space often filled with anxiety and despair about our possible futures. Only in our hearts can we exist in the here and now, with excitement, fear, and wonder.

In search of the ideal welcoming and ethical community and after days singing along with the car radio and nights sleeping alone in my bed roll, there was that sudden first glimpse of the beautiful blue of San Francisco Bay. Then, over the bridge, I-80 suddenly ended, spilling me into a crowded, hilly neighborhood.

My initial destination was the floor of my friend's apartment, offered as temporary refuge. For the rest of my young adulthood, I sought my ideal community, somewhere along the Urban Far-West Coast.

A decade later, sitting on the seawall in Pacific Beach, gulls and pelicans circling, I found myself listening to the waves rolling home. Maybe I'm part of a wave, I wondered, one of many waves of human immigrants, rolling across the continent, rolling the world.

And maybe soon we're all becoming immigrants, all in rolling waves of climate refugees, fleeing our global climate crisis, our global community crisis, our crisis of consciousness, fleeing ourselves? Perched on the edge of the continent, I realized there was no perfect place left on Earth to immigrate to. There probably never was.

Looking back, perhaps that was the day when I also gave up looking for the last remaining perfect urban climate and started working to create a better inner climate in my own body, mind, and spirit and, hopefully, working with other people to co-create new and better communities.

Economics, Politics, and the Climate Crisis

Is community something we find, something we create, or perhaps something of both? With so many of us now experiencing a kind of community hunger at a time when our nation is also being ravaged by a loneliness epidemic, we need a deepening consciousness about what kinds of communities in which we need to live and a commitment to cooperate with others in building them. We need to converse about what kind of nation we want to have and what kind of people we want to be.

And as we talk, we need to explore why there is such a perfect storm of crises happening now, all at the same time. There is the loneliness crisis. Climate crisis. Community hunger. Polarization crisis. Threats to democracy. Conspiracy theories. War and threats of war. Pandemic, the crisis of post-truth culture. Perhaps they are all happening now in large part because these and the other major crises are interconnected.

We will now examine climate-sensitive practice with local and global communities. We will explore how our macro-level economic and political policies are associated with the climate crisis and other converging crises and how some macro-level intervention strategies can be used to help to address the climate crisis.

Capitalism, Politics, and the Climate Crisis

Definitions and Introduction

There are many forms of democracy, as defined by experts in in the literature, and populations across the world also may have different views of what a democracy is or should be (e.g., Pew Research Center, 2021). In the United States, we currently live in what most experts call a ***liberal democracy***, which we can think of as a form of government in which, ideally, the people elect rulers who respect our individual rights and status equality, political parties "accept the legitimacy of defeat," and people and their rulers cooperate together for the highest good (Wolf, 2023, p. 5).

What is ***market capitalism***? We can think of it as an economic system in which production is privately owned and people's income is distributed through markets. There is an ongoing debate about how much market capitalism is based on human greed. What we can say is that greed is probably a significant factor in all political systems, because it is a part of human nature. On the other hand, we can also say that there are also altruism and compassion in human nature and that participants in every political system in the world can be motivated by both greed and altruism.

Wolf (2023) writes:

> Liberal democracy and market capitalism share a core value, a belief in the value and legitimacy of human agency, in political and economic life. ... Neither ... will function without a substantial degree of honesty, trustworthiness, self-restraint, truthfulness, and loyalty to shared political, legal, and other institutions. (p. 7)

We can agree with Wolf that, like all political and economic systems, both democracy and capitalism must be justly law-governed to be reliably effective.

We can think of ***politics*** as activities associated with governance of a country, especially the competition between people who seek power. Today, we see in the United States as well as in many other democracies *status anxiety*, which is concern by many that they are losing status to other people and a root cause of current populism politics. ***Populism*** politics have a hostility toward perceived elites and a rejection of pluralism (which is an appreciation of diversity) replaced by the emphasis upon the protection and enrichment of a particular subpopulation of people who may feel status anxiety but now want to view themselves as society's truly "authentic" people (i.e., those most deserving of status, power, and wealth; Wolf, 2023).

Thus, the current rise of populism in the United States and many other global democracies threatens to shift the cultural focus away from facing together our major crises and rather toward a competition between those who feel status anxiety and other groups whom they view as a threat to their status quo.

Finding the Radical Middle

In our text, we have shown how our climate crisis is associated with the ecobiopsychosocialspiritual well-being of human life and of all life on Earth. In this chapter, we will focus on how the climate crisis is associated with our economic and political environments, which is the *eco* part of the ecobiopsychosocialspiritual.

As we have done throughout the book, in this chapter, we will also strive to find the radical middle that exists between the extreme and dualistic positions that many people may take in our current

FIGURE 10.1 Examples of the Radical Middle Between Two Extreme Positions

economic and political divides. For example, many of us like to characterize people who vote differently from us as either economic "capitalists" or "socialists" or either political "conservatives" or "liberals." In Figure 10.1, we see illustrations of some of the extreme ways in which many of us view people whom we put into these kinds of categories. We also can see that there are positions between the extremes in the *radical middle* that we can take, which allows for the possibility of bridging these differences that currently divide us.

In addition to finding the radical middle, another strategy we can use to help heal the differences that divide us is to study the actual differences and similarities between us. In Table 10.1, for example, the key differences between capitalists and socialists are summarized. When we truly understand our differences and similarities, we stand a better chance of forming relationships, finding a common ground, and cooperating for the common good.

Indeed, for example, although the differences between our two major political parties are accentuated in many online conversations and media presentations, there is evidence that there is actually a growing agreement between the parties on some important issues. Areas of convergence include "putting Wall Street in its place," bringing "critical industries back to America," and resurrecting "an obligation to rebuild America's workforce" (*The Economist*, 2023a, p. 1).

TABLE 10.1 Key Differences Between Capitalism and Socialism

Socialism is an economic and political system under which the means of production are publicly owned. Production and consumer prices are controlled by the government to best meet the needs of the people. Capitalism is an economic system under which the means of production are privately owned. Production and consumer prices are based on a free-market system of supply and demand.
Socialism is most often criticized for its provision of social services programs that require high taxes and that may decelerate economic growth. Capitalism is most often criticized for its tendency to allow income inequality and stratification of socioeconomic classes.

Source: Longley, 2022.

Local Community

In *This Changes Everything: Capitalism vs. the Climate*, award-winning journalist Naomi Klein (2014) links our capitalistic economic system with both the climate crisis and the difficulties in our cultural and political worlds. She describes deregulated capitalism as the fundamental ideology of U.S. culture and attributes the lack of progress in our dealing with the climate crisis as linked to the difficulties involved in challenging this ideology, which is defended by the most powerful people and organizations in the United States (including our major polluters).

In Table 10.2, we list some of the intervention strategies that Klein suggests we make in within this country. Similar reforms could be expanded into many other countries in the world. As you can see, all these strategies are associated with current U.S. economic and political policies, and all would require standing up to powerful interests that currently benefit disproportionately from capitalism. As Klein points out, unfortunately, in our current system, lobbyists with the most money are the ones who are usually listened to.

For example, the first strategy, which involves installing universal health care, would have many potential benefits, including those that would help ameliorate the climate crisis (e.g., flood and wildfire assistance) and those that would benefit all people in the United States (e.g., lower health care costs and elimination of unnecessary illnesses and deaths). However, those who benefit most from the current insurance industry (e.g., owners of large insurance corporations) would be likely to continue to oppose any legislation that would upgrade our currently costly and perhaps unnecessary insurance industry (Klein, 2014).

TABLE 10.2 Some Policy Strategies That Can Help Ameliorate the Climate Crisis and Protect Populations From Negative Climate Impacts

Strategy	Details
Universal health care	Protect the poor and vulnerable from climate emergencies
Economic equality	May include universal income
Strengthened government	Protect us from disasters Regulate industries
Distribution of sacrifices equitably	Protect those who have to lose employment, homes, or income
Ensuring that powerful and wealthy classes pay fare share	Create fair tax structure
Regulation of corporations	Ensure they are reliable when they're required by law to be reliable
Rebuilding infrastructure	Promote public transit instead of more highways Implement smart grids Retrofit factories to produce green machinery
Planning for green job creation	Hold retraining and educational programs

(Continued)

TABLE 10.2 *(Continued)*

Strategy	Details
Planning for green power	Manufacturing, installation, service (such as electric charging stations for electric vehicles [EVs])
Creating climate-smart agriculture	Retrain agricultural workers and manufacture new equipment
Saying no to big oil and gas and yes to lower emissions	Gradually phase out these dangerous fuels
Reducing waste	For example, eliminate the current tragic amount of food waste in the United States

Source: Klein, 2014.

In our local communities, at the city, state, or even national level, there are many ways to work together to effectively deal with the climate crisis. A part of our work to make macro-level policy changes is to support the development of new green technologies. Technology by itself will not save us, and many proposed technological remedies to the climate crisis may have serious potential side effects (Klein, 2014). Nevertheless, there are emerging technologies that may all prove to contribute to the reduction of greenhouse gas (GHG) emissions and increased well-being of all life on the planet. *Fast Company* (Peters, 2023) recently published a list of 10 remarkable emerging U.S. technologies that are summarized in Figure 10.1. I added an 11th example, from Europe, that fascinates me (Brown, 2023).

TABLE 10.3 Examples of New Green Technologies

Green Technologies	Details
1. A gas utility works with clean geothermal heat.	Boston gas company Eversource is helping a neighborhood switch to geothermal heat.
2. A new factory produces carbon dioxide (CO_2)-based jet fuel.	A Washington state startup called Twelve plans to use captured CO_2, water, and renewable electricity to make jet fuel.
3. We are replacing gas stoves with cleaner alternatives.	Because of the health and climate risks from cooking with gas stoves, people are switching to using induction stoves.
4. The first commercial direct air capture in the United States has occurred.	In California's Central Valley, a new startup called Heirloom uses crushed mineral powder to pull CO_2 from the air.
5. Farms are using a new CO_2 removal solution.	A= startup called Lithos uses basalt dust on farms to increase yields and capture CO_2.

6. Zero-emissions ships are being designed.	A planned cruise ship will use sails covered in solar panels that use both wind and solar power to reduce emissions.
7. More options exist to produce clean heat in factories.	To replace gas and coal heating, a startup called Antora uses renewable blocks of carbon to heat factories.
8. The Massachusetts Institute of Technology's new tech turns concrete into a supercapacitor.	To reduce emissions from cement, researchers make cement with carbon black, creating blocks that store energy for homes.
9. We can now more cheaply retrofit inefficient buildings.	A startup called Hydronic Shell Technologies makes panels with built-in insulation and heating and cooling equipment.
10. New tech supports renewable energy.	A startup called Fervo drills down to deep underground hot rocks that heat water to create steam that generates electricity.
11. The first solar-powered truck reduces shipping emissions	Swedish vehicle manufacturer Scania has produced a solar-powered truck with solar panels on the trailer.

Sources: Peters, 2023; Brown, 2023.

Helping professionals can help educate community members about such innovations. We can also encourage our students, clients, and patients to consider supporting governmental policies that provide economic incentives for the development and use of green technologies as well as support the industries that create and distribute them.

Although the current Republican presidential candidates generally seem to discount the importance of addressing the climate crisis, there are conservative leaders who are concerned. For example, Mariannette Miller-Meeks, an Iowa Republican, wants to see GOP leaders become more involved in the climate crisis. However, she does not believe that fossil fuels should be phased out quickly and critiques Democratic climate legislation efforts that "take away choice" (New Yorker Radio Hour, 2023).

The National Academies of Sciences, Engineering, and Medicine (National Academies) recently published a remarkable report with more than 80 recommendations grading how we can best move the United States away from fossil fuel dependency by 2050 (Root, 2023). This report is the second of two offering recommendations regarding the United States' transition to a decarbonized energy system. The National Academies' work influenced the Infrastructure Investment and Jobs Act of 2021, CHIPS (Creating Helpful Incentives to Produce Semiconductors) and Science Act of 2022, and Inflation Reduction Act of 2022.

"The report addresses energy justice and equity, public health, the workforce, public engagement, clean electricity, the built environment, land use, transportation, industrial decarbonization, the financial sector, the future of fossil fuels, and state and local government roles" (National Academies, 2023, p. 1). Table 10.4 displays the key themes in this second report. In other words, a team of some of most talented scientists has already created a blueprint detailing how the United States can contribute directly to reducing GHGs and indirectly by modeling for other nations how to do the right thing for the well-being of all people.

TABLE 10.4 Key Themes From the National Academies Report on U.S. Greenhouse Emissions

Broadening the climate policy portfolio
Ensuring equity, justice, and health
Strengthening the U.S. electricity system
Supporting rigorous and transparent analysis and reporting
Ensuring procedural equity in infrastructure planning
Reforming financial markets
Building the needed workforce and capacity
Updating targets for the industrial and building sectors
Meeting research, development, and demonstration needs
Managing the future of the fossil fuel sector

Source: National Academies, 2023, p. 1.

Democracy and the Climate Crisis

In *The Great Experiment: Why Diverse Democracies Fall Apart and How They Can Endure*, Mounk (2022) offers a helpful analysis of what is happening in the United States today and what we can do about it. He reminds us that any multiethnic society will never be completely homogeneous, and, therefore, we need to develop shared life experiences and beliefs to build a more inclusive society together. For example, when people share nature together, the shared experience can bring us together. Public parks in both urban and rural areas can serve as shared experiences that are open to all.

Some of the differences that divide us include the slow pace of improvement in living standards, lack of economic progress in some ethnic and religious groups, lack of responsiveness in many of our institutions, and rise of cultural and political polarization. Online misinformation and disinformation from both domestic and foreign sources also contribute to the divides in our society.

We can, however, find a middle ground between the "dark histories" that different populations carry with them, says Mounk. We can emphasize the things we share rather than the differences that currently divide us and refuse to think of states as either red or blue. Other strategies that can help democracies find common ground include reforming the electoral system, reducing disparities in employment education, and moving forward with co-creating a real, equitable, and inclusive society.

Although there is no one simple way to keep our democracies strong, we professional helpers can remind ourselves and our communities that the well-being of society is ultimately more important than political victories.

Global Community

As we look at the climate crisis from a global perspective, the extent of climate justice issues stands out. As we reviewed in Chapter 8, researchers have identified associations between the extreme and still-growing inequities in wealth and power in our world and those regarding who contributes to and suffers from the climate crisis. A recent report published by *The Guardian*, international charity Oxfam, and the Stockholm Environment Institute explains that climate change and "extreme inequality" have become "interlaced, fused together and driving one another." In 2019, 16% of carbon emissions came from the top 1% wealthiest people, which is equivalent to the emissions produced by the 66% poorest people on Earth (about 5 billion people). The wealthiest 10% produced about half of the emissions of that year (Cohen, 2023, p. 1).

Chiara Liguori, Oxfam's senior climate justice policy adviser, described this inequity in another way:

> It would take about 1,500 years for someone in the bottom 99% to produce as much carbon as the richest billionaires do in a year. Using the "mortality cost of carbon" methodology, researchers estimate that the carbon dioxide produced by the top 1% in 2019, about 5.9 billion tons, would raise global temperature enough to contribute to the deaths of about 1.3 million people. (Cohen, 2023, p. 1)

We helping professionals can encourage people to commit to faith-based activism through advocacy as well as by serving themselves in government. For example, Chile is currently dealing with a 13-year drought. As Chilean minister for the environment, Maisa Rojas has helped create and enforce regulations on Chile's mining industry so that her country can accomplish carbon neutrality by 2050. Before assuming her present position, Rojas spent the past 15 years serving as a climate scientist and geophysics professor at the Universidad de Chile (Cardenas, 2023).

Dealing With Backlash and Other Resistance to Climate Solutions

In Europe, public concerns about the global economy and political stability as well as a related populist backlash have forced many nations to pull back on green goals, such as lowering emissions and promoting sustainable farming and greater biodiversity (Schwageri, 2023).

There are many reasons for political backlashes. Such factors include climate denial, fear of higher taxes, higher energy prices, costs of installation of new equipment, and concern that other countries are not doing their part. Populist politicians often build on these fears by using such tactics as exaggerating costs of a green future, lying about plans that the present government is making, and spreading culture war-related lies (such as "they will force you to eat tofu burgers").

What can governments do? To reduce public fear and backlash, governments can assume more of the unpleasant and expensive tasks, such as providing public heat pumps for an energy grid, providing realistic timeframes for changes to EVs, provide help with startup costs, and being honest about real challenges (Schwageri, 2023).

Ultimately, the wealthiest and most powerful nations cannot preach about sustainability to other nations if we don't lead by doing the right thing ourselves. Interestingly, climate change is a less divisive issue in most poor countries than in the wealthy ones. This is despite the fact that the poorer nations are often the ones most affected by the climate crisis while having the fewest resources to deal with climate disasters and pay for the costs involved in reducing GHG emissions (Lakhani, 2023). Fortunately, we are beginning to reduce those costs, and green energy and products are often already cheaper in many areas of our economy. As *The Economist* (2023b) summarized this year, despite the many challenges and risks involved in reducing carbon emissions, overall, the "most dangerous thing now is to go too slow."

Perhaps it is appropriate to include the following quote at the end of this chapter, from Klein, who wrote about the biggest obstacle to the climate solutions that most of us want to see implemented at the local and global community levels:

> So my mind keeps coming back to the question: what is wrong with us? What is really preventing us from putting out the fire that is threatening to burn down our collective house? I think the answer is far more simple than many have led us to believe: we have not done the things that are necessary to lower emissions because those things fundamentally conflict with deregulated capitalism, the reigning ideology for the entire period we have been struggling to find a way out of this crisis. We are stuck because the actions that would give us the best chance of averting catastrophe—and would benefit the vast majority—are extremely threatening to an elite minority that has a stranglehold over our economy, our political process, and most of our major media outlets. (Goodreads, 2024)

We have seen that the climate crisis can be seen in many ways. We can argue here that the climate crisis is also humanity's opportunity to develop our own self-empowerment, in which we simply recognize the authority and sovereignty of the average person who of course wants clean water, air, and land for their family.

QUESTIONS FOR REFLECTION

1. Many helping professionals feel less confident making interventions at the local and global community levels than at the family and institutional levels. Do you have similar feelings? Please explain. Why do you think you feel the way you do? What, if anything, do you want to do about the different levels of confidence you may feel about our work?
2. Would you like to see the United States become a world leader in clean energy? Why or why not?
3. What kinds of extreme and dualistic positions on economics and politics have you read or heard in the past month? What were your reactions? Can you find a radical middle to these positions?
4. What do you think about Klein's thesis that deregulated capitalism is a primary factor associated with the climate crisis? Please explain.
5. We have seen that although the majority of people in the United States express concern about the climate crisis, many politicians refuge to represent that view when passing legislation. How do you think and feel about that happening in a democracy such as ours? What kind of intervention strategy might help change this situation?

REFERENCES

Brown, H. (2023). *Could the world's first solar power truck be the answer to decarbonising haulage?* euronews, December 15, 2023. https://www.euronews.com/green/2023/12/15/could-the-worlds-first-solar-power-truck-be-the-answer-to-decarbonising-haulage

Cardenas, J. (2023). Maisa Rojas is putting political power behind climate science. *Vox*, November 29, 2023. https://www.vox.com/23945418/maisa-rojas-minister-environment-chile-future-perfect-50-2023

Cohen, L. (2023). *World's richest 1% emit as much carbon as 5 billion people, report says.* CBS News, November 21, 2023. https://www.cbsnews.com/news/worlds-richest-carbon-emissions-climate-change-report/?ftag=CNM-00-10aac3a

The Economist. (2023a). The American left and right loathe each other and agree on a lot. July 13, 2023. https://www.economist.com/united-states/2023/07/13/the-american-left-and-right-loathe-each-other-and-agree-on-a-lot?utm_id=1690933&sfmc_id=0033z00002rpJeVAAU

The Economist. (2023b). How to deal with the anti-climate backlash. October 14, 2023, pp. 10–11.

Goodreads. (2024). This changes everything *quotes.* https://www.goodreads.com/work/quotes/41247321-this-changes-everything-capitalism-vs-the-climate

Klein, N. (2014). *This changes everything: Capitalism vs. the climate.* Simon & Schuster.

Lakhani, N. (2023). Why loss and damage funds are key to climate justice for developing countries at COP28. *The Guardian*, November 29, 2023. https://www.theguardian.com/environment/2023/nov/29/why-loss-and-damage-funds-are-key-to-climate-justice-for-developing-countries-at-cop28?CMP=oth_b-aplnews_d-1

Longley, R. (2022). *Socialism vs. capitalism: What is the difference?* Thought.co, April 11, 2022. https://www.thoughtco.com/socialism-vs-capitalism-4768969

Mounk, Y. (2022). *The great experiment: Why diverse democracies fall apart and how they can endure.* Penguin Random House.

National Academies of Sciences, Engineering, and Medicine. (2023). *New report provides comprehensive plan to meet U.S. net-zero goals and ensure fair and equitable energy transition* [News release]. October 17, 2023. https://www.nationalacademies.org/news/2023/10/new-report-provides-comprehensive-plan-to-meet-u-s-net-zero-goals-and-ensure-fair-and-equitable-energy-transition

New Yorker Radio Hour. (2023). *Talking to conservatives about climate change.* August 18, 2023. https://www.newyorker.com/podcast/the-new-yorker-radio-hour/talking-to-conservatives-about-climate-change

Peters, A. (2023). 10 climate tech innovations that give us hope for 2024. *Fast Company*, December 15, 2023. https://www.fastcompany.com/90992666/10-climate-tech-innovations-that-give-us-hope-for-2024

Pew Research Center. (2021). *Democracy.* October 21, 2021. https://www.pewresearch.org/global/2021/10/21/spring-2021-democracy-appendix-a-classifying-democracies/

Root, T. (2023). *Scientists lay out a sweeping roadmap for transitioning the U.S. off fossil fuels.* Grist, October 17, 2023. https://grist.org/energy/us-scientists-lay-out-a-sweeping-roadmap-for-decarbonization/

Schwageri, C. (2023). Shifting political winds threaten progress on Europe's green goals. *Grist*, December 10, 2023. https://grist.org/energy/shifting-political-winds-threaten-progress-on-europes-green-goals/

Wolf, M. (2023). *The crisis of democratic capitalism.* Penguin.

SECTION 3

Climate Change in Ecosystems

IMG III.1

IMG III.2

As we mentioned in Sections 1 and 2, the main thesis of this text is that the roots of our climate crisis can be found in our own human consciousness, and as we change our consciousness, we can help heal and transform the climate of our interrelated human systems and ecosystems.

What is an ***ecosystem***? According to the National Geographic Society (2023, pp. 1–2):

> an ecosystem is a geographic area where plants, animals, and other organisms, as well as weather and landscape, work together to form a bubble of life . ... The whole surface of Earth is a series of connected ecosystems. Ecosystems are often connected in a larger biome. Biomes are large sections of land, sea, or atmosphere. Forests, ponds, reefs, and tundra are all types of biomes, for example.

In Section 3, we will examine our practice with ecosystems and the biomes they belong to, including the human biomes that we will call *anthromes* and that have been altered by sustained intensive interaction with humans. In the two previous images, we see examples of a biome and an anthrome.

We humans are part of our ecosystems. A climate change in an ecosystem occurs when we humans change our beliefs and behaviors regarding each other, other living things, and the environments that sustain their life. For example, if people in the United States decided to cooperate together to eliminate food waste, we would conserve 30%–40% of our total annual food supply, which is at least 133 billion pounds and $161 billion worth of food (U.S. Food and Drug Administration, 2023). The United States wastes more food than any other nation, and such a savings could have tremendous benefit to poor families and could also help lower greenhouse emissions significantly.

Climate change in ecosystems is about cooperation with other people, other living things, and the world around us. The Latin root *cooperationem* refers to "working together." In our text, we have examined many of the obstacles to cooperation, including our lack of consciousness about ego, our lack of values, knowledge, and skills in dialogue practice, and our lack of a sense of connection in a rapidly changing world.

In Section 1, which includes Chapters 1 through 5, we looked at how the state of our *human consciousness* is associated with our climate crisis and with potentially with desired climate change in human-systems and ecosystems. We showed how *increased consciousness*, which we defined as increased reverent awareness, can help humanity heal and transform ourselves and our world. In Section 2, we showed how *dialogue practice* can help us heal and transform humansystems so that humanity can better cooperate together to heal the ecosystems that support all life on Earth. We also introduced the concept and basic elements of climate-sensitive practice with humansystems.

In Section 3, we will explore how we can use both consciousness work and dialogue practice to help make a climate change in our relationships with our ecosystems so that we can support a healing of and transformation in the climate crisis. Chapter 11 provides an introduction to climate-sensitive practice with ecosystems, including engagement, assessment, and intervention strategies. In Chapter 12, we will introduce work that we can do in connecting with other living things and ecosystems. We have looked in our text at how we human beings are multidimensional ecobiopsychosocialspiritual beings, with multiple ways of knowing, which enables us to relate in many ways with the universe we live in. We will explore how we can connect with other living things and ecosystems through the elements of earth, wind, fire, and water. We will study practice with biomes in Chapter 13 and with anthromes in Chapter 14. Finally, in Chapter 15, we will take an inclusive look at consciousness and connection.

REFERENCES

National Geographic Society. (2023). *Ecosystem*. https://education.nationalgeographic.org/resource/ecosystem/

U.S. Food and Drug Administration. (2023). *Food loss and waste*. December 4, 2023. https://www.fda.gov/food/consumers/food-loss-and-waste

Credits

IMG 11.1

Inclusive Practice With Ecosystems

There must be a better way to make the things we want, a way that doesn't spoil the sky, or the rain, or the land.

—PAUL MCCARTNEY (GOOD GOOD GOOD, 2023)

The fall [of the Mayan civilization] has been laid at the door of overpopulation, overuse of natural resources, and drought. It is true that the Maya were numerous; archeologists agree that more people lived in the heartland in 800 A.D. than today. And the Maya had indeed stretched the meager productive capacity of their homeland. The evidence for drought is compelling too. … [But] the societal disintegration in the south was due to not to surpassing ecological limits [but to] political failure to find solutions.

—C. C. MANN (2011, PP. 316–318)

We sit in Maurice's living room as we wait for the program to begin. The house is still, except for nervous laughter as the doorbell announces more guests arriving. They take their places in the semicircle, facing the television set. Most eyes stare at the still-empty screen.

In November 1983, I was invited by my professor, Maurice Freidman, to join a small gathering at his house to watch the new film *The Day After*. The fictional movie gave audiences an experience of nuclear war, in part by following the experiences of several families in Lawrence, Kansas, and Kansas City, Missouri, which are two cities in the story that are bombed in a nuclear exchange between NATO (the North Atlantic Treaty Organization) and the Warsaw Pact (led by the Soviet Union).

Maurice was a kind and brilliant man who, as Martin Buber's biographer, introduced me and many other students to Buber's approach to dialogue. Maurice understood the wisdom of facing the awfulness and shock of nuclear war together with other people in a space in which we could all offer mutual support to each other. He skillfully created a welcoming space for all of us to belong to.

Perhaps our current climate crisis is even more horrific for us to watch. Instead of a sudden exchange of awesome nuclear destruction, we have seen the Earth increasingly devastated, especially over recent decades, by the everyday activities of humanity. The climate crises is like a slow motion version of *The Day After,* except that it is real and happens every day.

We watched the film on that evening in San Diego and then shared our emotions and thoughts. We faced some of the horrors of nuclear war together, and we all left Maurice's home as members of a new inclusive community.

That experience taught me that it is possible to bring a group of strangers together, in order to collectively face a challenging issue, raise our consciousness about ourselves and our world, join in dialogue, and begin building relationships with each other.

What is the current climate of our human relationship with our ecosystems? As we have explored throughout our text, human activity is the major contributor to the accelerated global warming; air, water, and soil pollution; and species extinction on our planet today. For most of us, if not for all of us living today, it is painful to see these conditions continue to get worse. Many of us use distraction, minimization, or denial to ease this pain. However, growing numbers of people across the planet are becoming willing to face the reality of this challenge, despite the local and global politics and economics that can make significant policy change difficult. Although we will not stop working for policy change, we do not necessarily need social policies to change before we change our own hearts, minds, and behaviors.

How do the principles of inclusive practice inform climate-sensitive practice with ecosystems? In this practice, we help develop individual and collective consciousness, inclusive local and global communities, and faith-based activism, all of which can support a much-needed climate change in our relationships with our ecosystems. Such a climate change includes healing our grief about what we have done to ourselves, each other, and our planet; transforming our denial that we are all connected with each other and with the ecosystems in which we live; and developing new attitudes and actions that can help restore their well-being.

Throughout the text, we have built upon the principles of inclusive practice as we create practice approaches that address the climate crisis. *Inclusive practice* is a methodology that gives helping professionals a broad and comprehensive range of knowledge, skills, and values for our climate crisis work (see Derezotes, 2023, for a more complete description of inclusive practice in social work). What follows is a summary of how inclusive practice principles apply to our climate-sensitive practice, when we address our relationship with our ecosystems.

The inclusive practitioner is both an artist and a scientist. We value both the artistic factors in our work with people as well as the use of the scientific method to help inform and assess our work, because both artistic and scientific factors are associated with positive outcomes (e.g., Duncan et al., 2010). Artistic factors include the helping relationship; the professional's intuition, empathy, loving kindness and compassion; and the client's readiness for and commitment to change. Scientific factors include the data that we have collected from both practice-based evidence and evidence-based practice.

The inclusive practitioner is committed to and helps develop inclusive communities. They are aware that our work with ecosystems has become politicized by many of our fellow citizens, including some who seek to continue to exploit the environment for profit or power as well as some well-meaning people who are simply concerned about the survival and well-being of their families and communities. The inclusive practitioner is happy to listen to others who disagree with them, knows that deep change happens through relationships, and seeks to develop civil relationships with everyone in their communities. We never insist that the climate crisis is the only crisis or even the most important crisis, because we realize, as Peter Frankopan (2023) and Naomi Klein (2014) showed, that our climate crisis in interconnected with the other major challenges of our era, including war, xenophobia, economic inequality, political polarization, and mass-casualty terror.

Inclusive practice challenges us to work at all four levels of practice: the micro, mezzo, macro, and eco. As illustrated in Figure 11.1, our relationships with ecosystems can be addressed on all four levels.

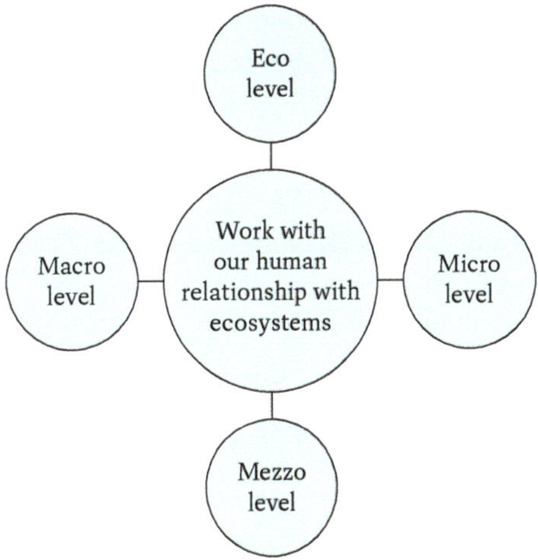

FIGURE 11.1 Work With Our Human Relationship With Ecosystems at the Four Levels of Practice

In Table 11.1, we illustrate examples of interventions at each level. Please note that entries in column 2 are not exclusive to one level; for example, consciousness work can happen at any of the four levels but will look different at each one. People can work on consciousness individually, in small groups, in larger community dialogues, or out in nature. Dialogue can also be used at all four levels, depending upon whether we want to use it with families, institutions, communities, or nature. The phrase *policy practice* includes collective work that usually builds on relationships developed in dialogue, in which people strive together to change local or global policies. The phrase *ecosystem interactions* refers to time spent outdoors in which individuals or groups participate in service projects and/or reflection.

TABLE 11.1 Examples of Practice Interventions in Work With Ecosystems

Level	Examples of Practice Strategies	Specific Examples of Interventions
Micro	Consciousness practice Dialogue practice	Working with clients individually or in families or groups to reflect on their emotions, thoughts, and behaviors related to nature
Mezzo	Consciousness practice Dialogue practice	Facilitating conversations in institutions and local communities, which may include periods of silent meditation and reflection
Macro	Conscious practice Dialogue practice Policy practice	Following up on dialogues to help organize communities to work together for policy change through protests
Eco	Consciousness practice Dialogue practice Ecosystem interactions	Helping establish projects in which volunteers care for local parks, gardens, streams, and beaches

Inclusive practice with ecosystems uses an ecobiopsychosocialspiritual approach in assessment and evaluation. In Table 11.2, we list examples of issues in our ecobiopsychosocialspiritual assessment of ecosystem practice. You will notice that in this table, all the examples involve examination of the relationships between human well-being and the well-being of our ecosystems. The phrase *ecosystem well-being* in the figure includes such factors as air, water, and soil quality; animal and plant species diversity; human interactions with wild and domesticated plants and animals; and access to parks and other more natural recreation areas.

TABLE 11.2 Assessing Ecobiopsychosocialspiritual Issues in Practice With Ecosystems

Dimension	Basic Focus	Some Assessment Issues in Practice With Ecosystems
Eco (ecological)	Our relationship with our human-made and "natural" environments	How is the well-being of our local and global humansystems related to the well-being of our local ecosystem? In what ways do my local and global communities act to either improve and/or deteriorate the well-being of these ecosystems? To what extent do my own behaviors serve to either improve and/or deteriorate ecosystem wellbeing?
Bio (biological)	Our relationship with our bodies	How is my physical wellbeing related to the wellbeing of my local ecosystem? How is my own physical self-care (or lack of care) associated with my care for local ecosystems?
Psycho (psychological)	Our relationship with our emotions and thoughts	How is my psychological well-being related to the well-being of my local ecosystem? To what extent does faith-based activism inform my life choices?
Social	Our relationship with other people	To what extent is my local community able and willing to face and agree on the science-based facts about the climate crisis? To what extent are the people in my local community able and willing to cooperate in addressing the climate crisis?
Spiritual	Our relationship with our own spirituality	What spiritual or religious beliefs, if any, was I taught about humanity's proper relationship with nature? What spiritual or religious beliefs, if any, were other people in my local and global communities taught? What is the relationship between my own spirituality and ecosystems?

Thus, when working with our clients, students, patients, and other community members, we invite them to join in assessing the many ways that their own well-being is interrelated with that of the ecosystems in which they live. Although we can examine statistical analyses of such relationships (e.g., the rates of respiratory illness in populations who live in areas with air pollution associated with weather inversions), we will also discover that there are always individual differences (e.g., some of us have asthma reactions to pollution, whereas others do not notice any reactions). There will also be individual differences in which developmental dimensions are noticed; for example, some people may primarily notice their psychological reactions to air pollution (e.g., feeling depressed), whereas others may especially notice physical reactions (e.g., feeling exhausted).

These assessment strategies can help clients become more conscious of their reactions to nature, which they may have likely had their whole lives but have either minimized or fully discounted. We will also find that, just like in the story about *The Day After* at the beginning of this chapter, some people may be more likely and more comfortable examining these reactions in a supportive group setting (rather than by themselves).

As inclusive practitioners, we also can use the seven paradigms of practice, regardless of our practice setting. For example, since everyone has a past, we can always ask people about their histories (psychodynamic paradigm) regardless of whether we work in a school, hospital, or counseling clinic. Examples of interventions with ecosystems drawn from each of the seven paradigms are shown in Table 11.3. Additional intervention strategies will be offered in subsequent chapters in Section 3. You will notice that the examples include consciousness raising (e.g., psychodynamic and transpersonal examples); service projects (cognitive behavioral, case management); social justice work (experiential); and dialogue (biopsychosocial, ecobiopsychosocial).

TABLE 11.3 Examples of Interventions With Ecosystems Drawn From Seven Paradigms of Practice

Paradigms	Examples of Interventions
Psychodynamic	Examine traditional familial and cultural patterns of interactions with nature that people were exposed to in childhood.
Cognitive-behavioral	Plan new green beliefs and behaviors, such as recycling household plastics, cardboard, and metals or putting food waste into a new backyard compost pile. Acquire an emotional and cognitive appreciation of the history of the life and Earth's ecosystems through study.
Experiential-phenomenological	Take groups of people out camping into nature who might not have had the opportunity or resources to do so.
Transpersonal	Have the client keep a journal in which they explore how they may have had spiritual experiences in nature, such as during hikes or fishing or hunting trips.
Case management	Refer people to an organization that uses volunteers to build new trails and help with erosion control in parks and forests.
Biopsychosocial	Take groups out for walks through parks or on trails as part of an ongoing dialogue group.
Ecobiopsychosocial (or deep ecology)	Co-create ongoing monthly community dialogues that address various issues in the climate crisis.

Source: Derezotes, 2023.

Inclusive Policy Practice Strategies

In the remainder of this chapter of this chapter, we will briefly introduce the four inclusive practice strategies we mention in Table 11.1: consciousness practice, dialogue practice, policy practice, and ecosystem interactions.

Consciousness Practice With Ecosystems

As we become more conscious of our attitudes and behaviors toward other living things and ecosystems, we can join in a transformation in the culture of humanity's relationship with ecosystems. In Table 11.4, we list examples of questions that can help guide consciousness work with ecosystems. When our students, clients, patients, and community members consider these kinds of questions, their deepened consciousness can lead to transformations in attitudes and behaviors.

Please notice how there are two questions on each line of the table. The "Explain" and "Why" follow-up questions ask people to do further and deeper self-exploration.

TABLE 11.4 Examples of Questions That Can Help Guide Consciousness Work With Ecosystems

Dimension	Question Helping Professionals Can Ask in Consciousness Practice
Eco (ecological)	Do I have any sacred location in the natural world? Explain. How do I feel when I am outside in a more natural settings? Why? How do I feel about the environment where I live? Why?
Bio (biological)	How does my body feel or act different during the different seasons? How does my body feel different when I am inside or outside? Explain.
Psycho (psychological)	Do I value humans over other living things? Why? Do I value humans over ecosystems? Why?
Social	Do I prefer natural settings when alone or with other people? Explain. Have I had a meaningful relationship with another living thing? Explain.
Spiritual	How connected do I feel with the universe? Why? When have I felt most connected with other living things? Explain. Do more natural settings have an influence on my own inner peace? Why?

Dialogue Practice With Ecosystems

Dialogue can utilize either "real-plays" or "role-plays" in the work. In real-plays, participants "play" themselves, and in role-plays, participants play other humans or other living things or ecosystems. In this section, we suggest that helping professionals use creative psychodrama approaches to dialogue, in which human participants can listen to and even give a voice to the plants, animals, and environments in their local environment. In such psychodrama, participants use spontaneous role playing to investigate and gain insight into their relationships with other living things and ecosystems. Psychodrama has a long history in the helping professions and draws from aspects of theater, role-play, and group dynamics.

There is considerable research supporting the usefulness of psychodrama techniques. After almost a century of psychodrama practice, authors of a review of the literature of psychodrama research found that "multiple studies have demonstrated the effectiveness of psychodrama and role-playing in producing therapeutic outcomes, by changing role expectations, as well as through improving interpersonal skills and role expansion. ... [P]sychodrama has since been adapted and practiced in more than a hundred countries across the globe" (Lim et al., 2021).

For example, as we have seen, one of the biggest divides in many communities is between those who are aware of and concerned about the climate crisis and those who are not. A dialogue facilitator can ask some participants to role-play the voices of various living things from the local environment and then have a psychodrama dialogue between the plants, humans, and other animals about how they are all feeling and thinking about global warming. Another example of a deep ecology exercise involves artwork; participants can be asked to draw a picture of their most sacred landscape and then share it with each other in a later processing discussion.

Policy Practice

Inclusive policy practice builds on consciousness and dialogue work. As the public becomes more conscious of the relationship we now have with ecosystems, more of us may start to want to change our attitudes and behaviors toward those ecosystems. And changes in attitude and behavior may lead to growing support for new green policies.

The goal of policy practice with humansystems is to promote social justice for all people (Cummins et al., 2023). In contrast, the goals of policy practice with ecosystems are to promote human consciousness of the interconnections we have with our ecosystems and promote and protect the well-being of those ecosystems, all through changes in local and global policies. Helping professionals can support policy change by devoting a fraction or the totality of our work toward that goal. And we can have an impact on policy change, even if we feel we only have values, skills, and knowledge at the micro and mezzo levels.

We do policy practice, for example, when we help our students, patients, clients, and other community members become more conscious of the issues in our local and global communities. We can ask them what issues matter most to them and how they are affected by those issues.

Helping professionals can also help plan and facilitate community dialogues in which people talk about the policy issues about which they are most concerned, perhaps with others who may disagree. When dialogue participants are interested, they may want to start to cooperate together in nonpartisan ways, meet to identify radical middle solutions, and begin cooperating to advocate policy change together. Helping professionals can attend such meetings and help organize such efforts.

Ecosystem Interactions

Professional helpers can also do policy practice by coordinating and leading group experiences outdoors and facilitating participation in service projects and/or reflection. Examples of urban outdoor events might include community gardening, community cleanup projects, and local tree planting.

QUESTIONS FOR REFLECTION

1. How is inclusive practice with ecosystems similar to inclusive practice with humansystems? How are they different?
2. In Table 11.1, we saw examples of practice interventions in work with ecosystems. Which of these examples are of most interest to you? Explain why.
3. Pick five questions in Table 11.2. Please respond to them in regard to how they apply to you and your own community.
4. Please review Table 11.3. With which of these paradigms are you the most comfortable leading? With which are you the least comfortable? Please explain.
5. Please also answer for yourself five questions in Table 11.4.

REFERENCES

Cummins, L. K., Byers, K. V., & Pedrick, L. (2023). *Defining policy practice in social work*. Routledge.

Derezotes, D. S. (2023). *Inclusive social work: A new vision of community practice*. Cognella.

Duncan, B. L., Miller, S. D., Wampold, B. E., & Hubble, M. A. (2010). *The heart and soul of change: Delivering what works in therapy* (2nd ed.). American Psychological Association.

Frankopan, P. (2023). *The Earth transformed: An untold story*. Knopf.

Good Good Good. (2023). *79 best sustainability quotes to inspire action*. October 2, 2023. https://www.goodgoodgood.co/articles/sustainability-quotes

Klein, N. (2014). *This changes everything: Capitalism vs. the climate*. Simon & Schuster.

Lim, M., Carollo, A., & Esposito, G. (2021). Surveying 80 years of psychodrama research: A scientometric review. *Frontiers in Psychiatry, 12*, 780542. https://doi.org/10.3389%2Ffpsyt.2021.780542

Mann, C. C. (2011). *1491: New revelations of the Americas before Columbus*. Vintage.

Credit

IMG 12.1

Earth, Wind, Fire, and Water

Some say the world will end in fire,
Some say in ice.
From what I've tasted of desire
I hold with those who favor fire.
But if it had to perish twice,
I think I know enough of hate
To say that for destruction ice
Is also great
And would suffice.
—ROBERT FROST (1923)

No matter where we come from, there is one language we can all speak and understand from birth, the language of the heart, love.
—IMANIA MARGRIA (GOODREADS, N.D.)

The first day I arrived, the first people I met in San Francisco told me they were magicians. They offered me a free tarot reading in recognition of my arrival in California in the spring of 1974. I remember sitting near a swing set on sunny grass in Golden Gate Park. The first card I drew was the "death" card, which they assured me did not necessarily mean that my actual death was immanent but that an ending and a new beginning in my life were now happening. They also told me that minor arcana cards in the deck are divided up into four suits—wands, swords, cups, and pentacles—which correspond to the elements fire, air, water, and earth.

Robert Frost, "Fire and Ice," New Hampshire, Henry Holt and Company, LLC, 1923.

Having always been a science geek, I knew that the idea that our universe is made up of the four elements—earth, air, fire, and water—was suggested at around 450 BC in ancient Greece. The Greek physician Hippocrates also taught that we humans are also made up of four *humors*, or temperaments that correspond to the four elements. Although there is little scientific evidence to support Hippocrates, the four elements themselves do align with the four states of matter of modern science: solid (Earth), gas (air), plasma (fire), and liquid (Home Science Tools, 2023).

There was something fascinating about the idea of the elements, perhaps partly because of their ancient origins, how they align with modern science, and they seem to correlate with symbols found in other wisdom traditions in the world. Western astrology, for example, divides the 12 signs into three fire signs (Aries, Leo, Sagittarius), three Earth signs (Taurus, Virgo, Capricorn), three air signs (Gemini, Libra, Aquarius), and three water signs (Cancer, Scorpio, and Pisces).

In the spirit of our inclusive practice approach, I decided to create a chapter in this text that draws from all ways of knowing and expressing, including the ways of logic, science, intuition, and imagination that academic neuroscientist and author Iain McGilchrist (2009) has written about. In her videos, the popular London astrologer Pam Gregory (https://www.youtube.com/c/PamGregory-Official/videos) often reminds her listeners that astrology is a "language of the heart" that can give many people an alternative way of communicating about life. Those of us who want to advocate for climate change in consciousness and in our relationships with humansystems and ecosystems need to be willing to use all these ways of knowing and communicating in order to reach all people in our local and global communities. The four elements provide a language of the body, mind, and spirit that appeals to many people.

Since I have always enjoyed the music group Earth, Wind & Fire, I decided to entitle Chapter 12 "Earth, Wind, Fire, and Water," which is a poetic title that substitutes the word *wind* for the word *air*. The chapter will provide exercises drawn from the language of the heart that helping professionals can use to encourage exploratory dialogues in which our clients, students, patients, and community members can raise their consciousness about our relationships with our ecosystems. There will be an introduction and four sections, one for each of the four elements.

Introduction

We can think of the four elements as both states of matter in the universe and basic orientations of human consciousness. These states and orientations are summarized in Table 12.1. The ***dimension of development*** refers to how each element corresponds to aspects of the ecobiopsychosocialspiritual model we have used in the text. The ***principles*** refer to what specific aspects are being developed toward the developmental ***goals***. The fifth column lists key ***relationships*** that each of us are evolving in our life; these are not necessarily only with other people, but also with parts of ourselves and our world. ***Life issues*** are areas of my life that I am working on, related to each of the elements.

TABLE 12.1 The Four Elements

Element	Dimension of Development	Principles	Goals	My Relationship With	Examples of Life Issues
Earth	Physical dimension	Stability, continuity	Patience, discipline	The Earth, my body	Material possessions, work, stubbornness
Wind (air)	Cognitive dimension	Perceiving, reasoning	Alertness, clarity	My mind, my thoughts	Understanding, stress, conflict
Fire	Spiritual dimension	Action, purpose	Will, courage	My spirituality, my creativity	Inspiration, burnout, enthusiasm
Water	Emotional dimension	Feeling, nurturing	Love, integrity	My heart, living things	Belonging, emoting, moodiness

Source: Forrest, 1988.

We can use this table when working individually or in groups and facilitate exercises in which we help people become more conscious of their relationships with the ecosystems in their world. In the rest of the chapter, we will explore various exercises we can use, some of which focus on other living things in our ecosystems and some that focus upon the elements of earth, air, fire, and water in our ecosystems.

The purpose of these exercises is to facilitate greater consciousness, or reverent awareness of my relationships with my world, so that I can live with an enhanced sense of connection with and kindness toward the ecosystems that support all life on Earth. These exercises are not about finding new reasons to feel more guilt or shame for ourselves, nor are they aimed at finding new ways to police each other's language or behaviors. Rather, the idea is that through increased awareness and acceptance of how we currently feel, think, and act toward our world, we can live in greater harmony and appreciation of our universe and of ourselves.

The exercises are written in the first-person point of view. In other words, the exercise questions are phrased as if I am asking the questions of myself. The helping professional can reorient the point of view of the questions as necessary when facilitating with the individuals or groups with whom they are working. Special attention is given in many of the exercises to distinguish between how I might feel, think, or behave in relationship with the topic being explored. Each exercise is preceded by a short introductory paragraph, and before using an exercise, helping professionals might want to first read or paraphrase that paragraph.

IMG 12.2

Earth

Earth

We call our home planet *Earth*. The name has ancient and uncertain roots: "Like many names of solar system objects, Earth's original name is long lost to history. But linguistics provide a few clues. *Ertha* [emphasis added] is an approximate spelling for 'the ground' (meaning, the ground upon which we stand) in Anglo-Saxon, one of many ancestor languages to English" (Howell, 2022).

EXERCISES ABOUT EARTH

1. What are my associations with the word *Earth*? When I hear the word, for example, do I first think of our planet, the soil under my feet, psychological grounding, or all of these and perhaps other associations?
2. How do I feel about these associations? To illustrate, I might have a feeling of awe and reverence for the Earth or perhaps a feeling of sadness about how it has become polluted and disturbed by human activity.

Dirt

We commonly use the word *dirt* to describe soil, as well as other things. According to the Online Etymology Dictionary (https://www.etymonline.com/), the word is a "metathesis of Middle English *drit, drytt* ... 'excrement, dung, feces, any foul or filthy substance' also 'mud, earth,' especially 'loose earth.'"

The word *dirt* is used to describe many things that are either considered foul or filthy or relate to "loose earth." I have a dirt bike in my garage for trail rides, for example; also, a product might be "dirt" cheap, and people might tell a "dirty" joke.

EXERCISES ABOUT DIRT

1. In what ways do I tend to use the word *dirt*? How do these uses reflect my feelings and beliefs about dirt? For example, do I associate *dirty* with *foul* or *filthy*?
2. How do my feelings and beliefs about dirt seem associated with how I feel, think, and act toward the soil that nurtures so much of life on Earth?

Humility and *Humis*

I am fascinated that the word *humility* seems to be derived from the word *humis*. According to the Online Etymology Dictionary, *humility* is the "'quality of being humble,' from Old French *umelite* '*humility*, modesty, sweetness' ... from Latin *humilitatem* ... 'lowness, small stature; insignificance; baseness, littleness of mind,' in Church Latin 'meekness'" from *humilis* ... 'lowly, humble,' literally 'on the ground,' from humus 'earth.'" Once again we see that the words we use suggest how we feel and think about our world; in this case, the etymology seems to suggest that we feel and think that the ground and soil is insignificant and lowly.

EXERCISES ABOUT HUMILITY

1. How do I feel and think about the words *humility* and *humiliation*? How do I see them as equivalent or different?
2. How do my answers to the previous question suggest how I feel, think, and act toward the Earth or the soil that sustains so much life on our planet?

Childhood Experiences

On way to bring people together is to ask them to share stories from their past, including about their early childhood. I have found that most adults like to share such stories, and in doing so, they frequently rediscover how they feel and think about their ecosystems.

EXERCISES ABOUT CHILDHOOD EXPERIENCES

1. What are my earliest and/or early childhood experiences regarding the Earth? For example, do I remember playing in the dirt, getting dirty, or even eating dirt?
2. How did my parents, other authority figures, or other people teach me about the Earth? For example, as a child, was I yelled at for playing in the dirt, getting dirty, or eating dirt?

Sacred Earthscapes

When we say that parts of nature are sacred to us, we are expressing that we highly value them and may offer them our respect, reverence, awe, and protection. When I ask people about their sacred earthscapes, they often identify such landscapes as mountains, mesas, canyons, valleys, meadows, and beaches.

EXERCISES ABOUT SACRED EARTHSCAPES

1. Do I have my own sacred earthscapes? If so, what are they?
2. How and why did these forms become sacred to me?
3. How do I feel, think, and behave when I am at my sacred earthscape or even just imagining being there?

Ground-Dwelling Life

There are many creatures and plant species that live in the ground, all over the world. In North America, these include moles, foxes, badgers, prairie dogs, groundhogs, worms, bacteria, borrowing owls, picas, and numerous insects and spiders. In addition, there are many species of mosses, ferns, cacti, trees, bushes, mushrooms, and grasses that are rooted in the ground.

When we say that we especially value or have an affinity with a living thing, we affirm that there are qualities in that living thing that we want to have and develop. There may be a number of reasons why a particular animal does not appeal to us, including the possibility that it has qualities that we also dislike in ourselves.

Readers can find many sources for exploring the wonderful world of "spirit" or "power" animals and can extend their world further by exploring plant life as well. For example, hikers and campers might take a copy of Ted Andrews's beautiful *Animal-Speak: The Spiritual and Magical Powers of Creatures Great and Small* (2014) with them and read about some traditional views of the wild creatures they encounter outdoors.

Professional helpers can help guide this exploration of animal and plant life. Many of my students who go on solo "vision quests" into the outdoors get frustrated as they wait for a suitable spirit animal to appear. They may report disappointment that they "only" saw a dragonfly at their campsite, until they read how dragonflies were seen by some Indigenous people across the world: "Dragonfly can help you to see through your illusions and thus allow your own light to shine forth. Dragonfly brings the brightness of transformation and the wonder of colorful new vision" (Andrews, 2014, p. 342).

We also want to honor our reverence for and affinities with plant life. Jane Struthers, for example, created her wonderful guide *Using the Wisdom of Trees Oracle: Oracle Cards for Wisdom and Guidance* (2017), which provides artwork, background information, and ancient history about some of the world's most dramatic trees. For instance, she writes that oak trees have long been associated with steadfastness, physical power, and inner emotional strength.

EXERCISES ABOUT GROUND-DWELLING LIFE

1. What kinds of living things do I most notice that dwell in the ground in my own local ecosystems?
2. Which of these living things do I value the most or have a strong affinity with? Why?
3. Which of these living things do I value the least or not have an affinity with?

Wind (Air)

IMG 12.3

Air

The Online Etymology Dictionary (https://www.etymonline.com/) describes the origins of our word *air*:

> c. 1300, "invisible gases that surround the earth," from Old French *air* "atmosphere, breeze, weather" (12c.), from Latin *aer* "air, lower atmosphere, sky," from Greek *aēr* ... "mist, haze, clouds," later "atmosphere" (perhaps related to *aenai* "to blow, breathe"), which is of unknown origin. ... To be **in the air** "in general awareness" is from 1875; **up in the air** "uncertain, doubtful" is from 1752. To build

> **castles in the air** "entertain visionary schemes that have no practical foundation" is from 1590s (in 17c. English had airmonger "one preoccupied with visionary projects"). Broadcasting sense (as in **on the air, airplay**) is by 1927. To **give (someone) the air** "dismiss" is from 1900. [emphases added]

When you study the previous etymology information, you see that there is a long-term tendency in our culture to associate the word *air* with such concepts as uncertainty, impracticality, or dismissiveness. When people today, for example, call someone an airhead, we are typically not complementing them, but rather implying that they that may have a lack of focus or even a lack of intelligence.

EXERCISES ABOUT AIR

1. How do I feel and think about air?
2. When am I most aware of the air in our sky?
3. To what extent do I use the word *air* to refer to uncertainty, impracticality, or dismissiveness?

Wind

Wind, of course, is moving air, and air is usually difficult for we humans to see directly. When air moves, we are more likely to notice it. Although our atmosphere is usually invisible to human eyesight, we can feel the air when there is wind, and we can often see the effects of wind in the movement of trees and grass, our own hair and clothing, and other things that can bend or move. Wind can sometimes feel wonderful; for example, when a cool lake breeze in the hot summer afternoon cools our bodies. Wind can also be frightening, such as when a tornado or hurricane starts to break tree branches and damage homes.

EXERCISES ABOUT WIND

1. What kinds of experiences have I had with wind?
2. What is the range of feelings and thoughts I have had about wind?
3. Have there been times I was frightened by strong wind? Have there been times when wind was soothing or enjoyable?

Inspiration

We can feel air when we breathe, when we inspire, but we usually tend to not to notice the sensation. The words *inspiration*, *inspired*, and *spirituality* all evolved from the Latin root *inspirare*, which means "to breathe." *Spirituality* thus refers to "the breath of life."

Plants breathe, too. Plants have tiny stomata, or mouths, which bring carbon dioxide into them and expel water so that photosynthesis can take place. I remember once walking with a friend through a cornfield on an early summer day in Iowa. My friend told me he could see the corn growing and feel the corn breathing. Responding to his language of the heart, I found I was also able to experience the corn plants growing and breathing.

EXERCISES ABOUT INSPIRATION

1. Can I pay attention to my breathing? Are my breaths more shallow or deep? Do I know why?
2. What is it like to live in and breathe in polluted air?
3. Why do I suspect that some ancient people equated spirituality with breath? How might I relate to their metaphor?

Flying Beings

There are flying beings that are part of most ecosystems on the Earth. Birds and insects are often the most noticeable animals in the air, but there are also many insects as well as mammals, such as bats, some marsupials, and flying squirrels that can also be seen flying, gliding, or parachuting across the sky.

EXERCISES ABOUT BEINGS WHO LIVE IN THE AIR

1. What kinds of living things do I most notice in the air, in my own local ecosystems?
2. Which of these living things do I value the most or have a strong affinity with? Why?
3. Which of these living things do I value the least or not have an affinity with?

Fire

Fire

IMG 12.4

Was fire considered by ancient people to be a form of energy or more a substance like earth and air? The Online Etymology Dictionary (https://www.etymonline.com/) tells us that the Proto-Indo-European (PIE) languages apparently had two roots for fire: *paewr-* and *egni-* (source of Latin *ignis*). The former was *inanimate*, referring to fire as a substance, and the latter was *animate*, referring to it as a living force (compare *water* [n.1]).Today we may fire a gun or rocket, be fired from work, feel fired up about a football game, sing a song about light my fire, or say that someone has a fiery (angry) personality. Sadly, we also hear more frequently about destructive wildfires. In 2023, we had some of the most extensive and damaging fires ever recorded in locations across the globe.

EXERCISES ABOUT FIRE

1. What are my associations with the word *fire*?
2. How do I feel about these associations? Do I tend to feel positive about fire or more negative? Why?
3. After I read Robert Frost's poem "Fire and Ice" at the beginning of this chapter, how do I answer his question about whether humanity is more likely to perish from fire or from ice? Why?

Climate and Fire

We all know that the average temperature of the planet is rising quickly; however, temperature extremes have also become more likely across the planet. These extremes can vary across time and location, as human-caused climate change interacts with such factors as geography, seasonal changes, and natural weather cycles. Thus, in the middle of our global warming, some locations on the planet can have record-breaking cold-weather events.

EXERCISES ABOUT WEATHER EXTREMES

1. Do I happen to live in an ecosystem in which extremes of heat are more common or extremes of cold?
2. What kind of weather extremes are the most difficult for me emotionally and/or physically? Drought or flood? Heat or cold? Please explain.

Climate Migration

Some of the most destructive and lethal climate emergencies are drought, flood, and wildfire. We have discussed in our text how more people are now considering the risk of such climate challenges as they try to plan where they would like to live. If they are fortunate to have the resources and opportunities to move, they might try to move to a safer location.

EXERCISES ABOUT CLIMATE MIGRATION

1. How important is the risk of drought, flood, and wildfire in the factors that influence where I want to live?
2. What kind of climate patterns are the most difficult for me emotionally and/or physically? Such patterns might be associated with the extinction of local wildlife, for example, or perhaps the loss of green spaces. Other issues might include the inability to go outside for long periods of time during heat waves, or the cost of air conditioning during such heat waves. Please explain.

Desert Creatures

There are creatures that benefit from fire. For example, low-severity fires help maintain open spaces between ponderosa trees, and higher-severity fires open up the canopy for lodgepole pines to thrive. Black-backed woodpeckers nest in trees killed by severe fires and like to dine on the beetles and other insects that feast on burned trees.

Although few, if any, beings appear to actually thrive inside a live fire, we can think about living things that thrive in the desert as life that has adapted to the fiery heat of the sun. Many desert plants are succulents that store water in their leaves, stems, fruits, and roots. Desert animals include specially adapted insects, reptiles, birds, and mammals called *xerocoles.*

EXERCISES ABOUT BEINGS WHO LIVE IN THE DESERT

1. With what kinds of living things am I most familiar that live in the desert?
2. Which of these living things do I value the most or have a strong affinity with? Why?
3. Which of these living things do I value the least or not have an affinity with?

IMG 12.5

Water

As was the case with the element of fire, water was also considered by ancient people to be both a form of energy as well as a substance. The Online Etymology Dictionary (https://www.etymonline.com/) tells us that the PEI languages "had two root words for water: **ap-* and **wed-*. The first (preserved in Sanskrit *apah* as well as *Punjab* and *julep*) was 'animate,' referring to water as a living force; the latter referred to it as an inanimate substance."

As human populations continue to increase, our water consumption rises, and water resources dry up in many locations, we read more often about concerns about water shortages. The previous image shows a desert river. The U.S. Colorado River, for example, provides water for seven thirsty states that are all suffering from various levels of drought and growing competition for shrinking water resources.

EXERCISES ABOUT WATER

1. We have learned that both water and fire have been considered in the past to be both *animate* and *inanimate*. How do I feel and think about these ideas? Do I also see these elements as having traits of both as well? Please explain.
2. How else do I feel and think about water?

Wilder Water

We could say that there is domesticated and "wilder" water. *Domesticated water* could be defined as existing in human-made containers, including tanks, canals, and impoundments. *Wilder water* lives in natural spaces such as lakes, rivers, glaciers, clouds, and seas. Unfortunately, as we have seen, much of our wilder water, increasingly contaminated by human-made pollutants, is becoming less wild every year.

EXERCISES ABOUT WILDER WATER

1. Do I feel and think differently about domesticated and wilder water? For example, what is my reaction when I see raw sewage being dumped into a public stream or bay? Please explain why.
2. Does it bother me when a wilder river is damned up or the shores of a wilder lake is built up into homes and condominiums? Please explain.

Blue Mind

In his book *Blue Mind* (2014), Wallace J. Nichols summarizes research that indicates human beings benefit in many ways by being near bodies of water. This research has found evidence that proximity to water has both psychological and physical benefits for many humans. Perhaps this affinity with water is partly related to the history of life on Earth, because all life on our planet is thought to have evolved from early beings who lived in our ancient oceans.

EXERCISES ABOUT BLUE MIND

1. Do I find that I have an affinity with water? Please explain.
2. What kind of psychological and physical reactions do I notice when I am near bodies of water?

Beings in Water

Life exists in both our freshwater and saltwater spaces. Freshwater exists not only in surface water spaces such as lakes and streams but also in ground water, ice, and glaciers. Saltwater exists in the oceans as well as in some lakes, ponds, underground spaces, and coastal saltwater intrusions.

Aquatic plants, also called *macrophytes* or *hydrophytes*, live in both saltwater and freshwater. Aquatic animals in saltwater and freshwater spaces include fish, worms, mollusks, reptiles, amphibians, mammals, and birds. Just as we explored our affinities with beings that live on the Earth and in the air, we can also discover our affinities with various forms of life in our watery spaces.

EXERCISES ABOUT BEINGS WHO LIVE IN THE WATER

1. What kinds of living things do I most notice in the water in my own local ecosystems?
2. Which of these living things do I value the most or have a strong affinity with? Why?
3. Which of these living things do I value the least or not have an affinity with?

QUESTIONS FOR REFLECTION

1. Please explore and respond yourself to questions from each of the four chapter sections for yourself.
2. What questions were most interesting and helpful? What additional questions would you like to add, to use personally and with the people in your community?

REFERENCES

Andrews, T. (2014). *Animal-speak: The spiritual and magical powers of creatures great and small.* Llewellyn Publications.

Forrest, S. (1988). *The inner sky.* Bantam.

Frost, R. (1923). *New Hampshire.* Henry Holt.

Goodreads. (n.d.). *Language of the heart quotes.* https://www.goodreads.com/quotes/tag/language-of-the-heart

Home Science Tools. (2023). https://learning-center.homesciencetools.com/article/four-elements-science/

Howell, E. (2022). *How did Earth get its name?* Live Science, May 31, 2022. https://www.livescience.com/32274-how-did-earth-get-its-name.html

McGilchrist, I. (2009). *The master and his emissary: The divided brain and the making of the Western world.* Yale University Press.

Nichols, W. J. (2014). *Blue mind: The surprising science that shows how being near, in, on, or under water can make you happier, healthier, more connected, and better at what you do.* Little Brown & Company.

The Online Etymology Dictionary (https://www.etymonline.com/).

Struthers, J. (2017). *Using the wisdom of trees oracle: Oracle cards for wisdom and guidance.* Watkins.

Credits

CHAPTER 13

IMG 13.1

Practice With Biomes

Sheer cliffs tower above sparkling waves. Smooth domes swell up before you on a portage. Craggy points reach out to caress the water. Oddly-shaped boulders loom in the depths or stand wet-footed in the shallows. They cradle the lakes, frame the vistas, and create stairs, rooms, and seats in your campsite home. … The tangible geology of the Boundary Waters begins about 2.7 billion years ago. The rocks of this era are part of the Canadian Shield, the heart of the continent. This "craton" is the base on which the rest of North America was formed. In the Boundary Waters, the shield is composed of three types of rocks: metamorphosed volcanic rocks, metasedimentary rocks, and granitic rocks.

—EMILY STONE (2020)

We now live in the Anthropocene—a geological epoch defined and dominated by the deeds of our species. We have changed the climate and acidified the oceans by releasing titanic amounts of greenhouse gases. We have shuffled wildlife across continents, replacing indigenous species with invasive ones. We have instituted what some scientists have called an era of "biological annihilation": comparable to the five great extinction events of prehistory. … Instead of stepping into the Umwelten *of other animals, we have forced them to live in ours by barraging them with stimuli of our own making. We have filled the night with light, the silence with noise, and the soil and water with unfamiliar molecules. We distracted animals from what they actually need to sense, drowned out the cues they depend upon, and lured them, like moths to a flame, into sensory traps.*

—ED YONG (2022, P. 336)

It was 1968, and we were six high school kids from Chicago driving north. After a few hours, we finally left the factories, towers, and freeways of the city. Then the village farmlands of Northern Illinois and Southern Wisconsin appeared and gradually yielded to longer stretches of pine and aspen forest, with peeks of occasional sky-blue lakes and rocky streams between the trees.

It was the summer before our senior year. We were starved for room to wander, explore, laugh out loud, and test our muscles. It was time to visit Mother Nature for the healing we needed. She did not disappoint us.

You can tell when you are getting closer to the Canoe Boundary Waters. Far Northern Minnesota is unmistakable. You start seeing more exposed rock formations, and there seems to be more water than land. There were no billboards, at least back then. Just water, rock, and forest as far as you can see—blue, light gray, and green. We had a week of adventures ahead of us. A bear growling outside of our tents at night; big, hungry pike right next to the shoreline; long portages; ravenous mosquitoes; a windy storm; full of stars.

Since moving across the country, I have not returned to visit the Boundary Waters for more than 40 years. I wonder what it is like these days, after reading that the annual visitation rate is now more than 165,000 people per year.

What happens to the experience of being there when you share the lakes and streams with more and more people? What happens to the wildlife that live there when more humans visit every year? How do we manage those wonderful areas in a way that is fair and just to the waters, forests, wildlife, and human visitors?

We might call the Canoe Boundary Waters a *biome*, which is a large section of land, sea, or atmosphere that remains relatively unaltered by human activity.

In this chapter, we will explore how we can protect living things, especially animal and plant wildlife in our biomes, from the conditions of our climate crisis. We will use the climate-sensitive practice schema of micro, mezzo, macro, and eco levels again. However, instead of being organized by these four levels of practice, this chapter is primarily organized by the practice topics that we will address. Then, organized *under* each topic heading, as illustrated in Figure 13.1, we will also explore some examples of micro-, mezzo-, macro-, and eco-level intervention strategies that can be used by helping professionals as they address that particular topic.

The practice topics all describe interrelated conditions with shared etiologies, as we discussed in earlier chapters. For example, the burning of fossil fuels is associated with many conditions associated with the climate crisis, including global warming, pollution, habitat destruction, loss of species diversity and species extinction, and climate justice issues. Examples of practice issues associated with each chapter topic are listed in Table 13.1. Figure 13.1 illustrates the four levels of practice, and Table 13.2 lists examples of practice strategies, interventions, and measurable goals that can be found in each of these four practice levels.

Climate-sensitive practice, introduced in Section 2 of this text, may use any of the practice approaches described in this chapter.

TABLE 13.1 Examples of Practice Issues Under Chapter Practice Topics

Practice Topics	Examples of Practice Issues
Global warming	Air pollution
	Species migration
	Rising sea levels
	Biodiversity loss
	Range shifts
	Invasive species
	Wildlife health
	Increased erosion
	Decreased harvest

Pollution	Air pollution Water pollution Soil pollution Light and noise pollution
Marine habitat and land habitat destruction	Rewilding Acidification Water temperature Marine heatwaves Deforestation
Loss of species diversity and species extinctions	Invasive species
Vacations	"Getting away from it all"

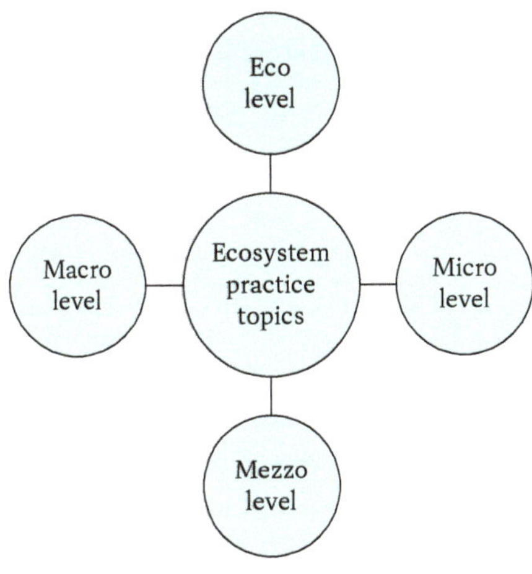

FIGURE 13.1 Work With Ecosystems at the Four Levels of Practice

TABLE 13.2 Examples of Practice Interventions in Work With Biomes

Level	Examples of Inclusive Practice Strategies	Examples of Interventions	Examples of Measurable Outcomes
Micro	Consciousness practice Dialogue practice	Co-therapy with nature Creation of new behaviors	Volunteering to maintain park trails Recycling of household waste
Mezzo	Consciousness practice Dialogue practice	Creation of new institutional and community behaviors	Community gardens Green buildings Xeriscaping Tree planting
Macro	Conscious practice Dialogue practice Policy practice	Creation of new social policies	Dialogues in all high schools Regulation of major polluters Enforcement of regulations
Eco	Consciousness practice Dialogue practice Ecosystem interactions	Community education Fund biome protection Rewilding projects	Healthy ecosystems Diverse wildlife populations Wildlife corridors

Faith-Based Activism

In this chapter, we will again face together some unhappy and probably painful realities that are occurring today in our local and global biomes. But as we discussed, there is every reason for continuing to embrace faith-based activism.

As we have discussed, I find that my students like to hear good news about the environment as well as updates on the climate crisis. I like to offer them both positive and negative science-based facts about what is going on. For example, according to naturalist Adam Welz:

> Human society is completely reliant on the predictable functioning of ecosystems. We are pushing those ecosystems into unstable states, driving up uncertainty, pushing ourselves deeper and deeper into the unknown. ... It will take lots of projects, big and small, all over the world, to help ecosystems begin to heal, but that effort could also ease the impact of climate-fueled events like the more intense fires, droughts and heat waves that are already reshaping our human lives. (Mann, 2023)

National Geographic (Mulvaney, 2024) recently published a report that included both disappointing and encouraging environmental news. On the one hand, in 2023, the global temperature kept rising, the pace of wildfires increased, the Artic and Antarctic both saw increased melting, and additional species extinctions were noted. On the other hand, we also saw the increased production and use of electric vehicles, an increase in renewable energy capacity, and the rediscovery of some species that were previously thought to be extinct.

Helping professionals can also help promote faith-based activism by serving to educate our community members about the climate crisis and encouraging them to face our reactions to what we are learning. As conditions change and we learn more, new language is invented to describe our changing world. In Table 13.3, for example, we include the "top 10 terms that capture the climate in 2023" (Yoder, 2023) from a *Mother Jones* article published at the end of 2023.

TABLE 13.3 Terms and Descriptions From *Mother Jones* Article

Term	Description
Air Quality Index	A color-coded measure of how dangerous the air is to breathe
Carbon insetting	Business-speak for companies reducing emissions in their own supply chains; an alternative to carbon offsetting
Climate quitters	People who resign from their jobs over concerns about climate change
Deinfluencers	Social media influencers who push back against extreme consumption
El Niño	A global weather pattern characterized by warmer-than-average temperatures in many regions
Global boiling	Like global warming, but much more worrying
Greenhushing	When companies go quiet on their environmental commitments
Noctalgia	The feeling of missing a dark night sky

Racketeer Influenced and Corrupt Organizations (RICO) Act	A law originally created specifically for the Mafia and organized crime that is now being applied to oil companies
White hydrogen	Naturally occurring hydrogen found underground

Source: Yoder, 2023.

We can also show our students, patients, and clients that there is evidence of advancement in some areas of concern and help them focus their activism on the key sectors that still need to change the most. For example, we know that in most countries, at least some progress is already being made to reduce the burning of fossil fuels. However, the aviation, shipping, and industry sectors are notably lagging in change around the world (Milman, 2024).

Other encouraging news can be found in recent polling results, covered by CBS News (Potts et al., 2024, p. 1):

> An increasing majority of people are now willing to face the realities of our climate crisis: "Polling shows that Americans' doubts about whether climate change is real have diminished. Most Americans think climate change is happening (72%) and is mostly caused by humans (58%), according to the latest Yale Program on Climate Change Communication quarterly survey from the fall. A slight majority of 53% understand that "most scientists think global warming is happening." The share of Americans holding each of these opinions has slowly increased over the last 15 years, but been fairly stable in the last few years, according to the Yale survey. Those numbers also vary by state: In California, 77% of residents surveyed think global warming is happening.

Global Warming

The accelerated global warming largely associated with human activity is having a profound negative impact on our biomes. According to the U.S. Geological Survey (2023), climate change has many impacts on plants, animals, and ecosystems, including unhealthy air pollution, species migration, rising sea levels, biodiversity loss, range shifts, invasive species, wildlife disease, increased erosion, and decreased harvest. The World Wildlife Federation (WWF; 2018) reported that a study it participated in found that "up to half of plant and animal species in the world's most naturally biologically rich areas, such as the Amazon and the Galapagos, could face local extinction by the turn of the century due to climate change if carbon emissions continue to rise unchecked."

How can helping professionals assist our community members in learning more about global warming and become faith-based activists? A simple strategy at the micro and mezzo levels is to move our work locations outdoors when possible. For example, *The New York Times* (Caron, 2024) recently published an article about how mental health practitioners are increasingly taking their clients on hiking, camping, and other outdoor adventures. Clients often report that opportunities to do therapy outdoors can help them connect more with nature and find an appreciation for all that our ecosystems give us. Educators, health care professionals, and other helpers can also provide similar experiences. I have frequently found that most of my students and clients enjoy opportunities to work outdoors.

At the mezzo and macro levels, we can continue to educate our communities about climate justice issues and encourage their activism around these issues. According to Accius (2024, p. 1):

> Extreme weather takes a heavier toll on children and older adults, and certain communities, including those that are under-resourced, rural, and communities of color. When we look at prevention and preparedness, we must consider those most at risk from extreme weather and environmental impacts if we want to address these challenges.

At the macro level, we can advocate for effective carbon credit systems. In a recent euronews article, Maria Mendiluce (2024) reported:

> The carbon finance community recently welcomed the launch of a new code of practice to rebuild trust in "high-integrity" carbon credits, designed to help governments and businesses (or even individuals) accelerate their transition to "net zero" emissions. Released by the Voluntary Carbon Markets Integrity Initiative (VCMI), this additional guidance enables buyers to make claims more credibly about their use of high-quality carbon credits. Simultaneously, the Integrity Council for the Voluntary Carbon Market (ICVCM) is addressing the supply of high-quality carbon credits by setting rigorous thresholds around disclosure and sustainable development. ... External accountability is essential and welcome, and new efforts like those of VCMI and ICVCM will help differentiate and validate in the market more robust claims and credits respectively and accelerate climate action. (pp. 1–2)

Pollution

Air Pollution

We also know that air pollution harms wildlife in our biomes. The effects of air pollution include loss of biodiversity, higher mortality risk, and overpopulation of certain species. Scientists have found that air pollution harms a wide range of animals, including pets, cattle, birds, wild animals, and insects. Respiratory issues, cancer, and changes in migration patterns are only some of the consequences (Pengue, 2022).

At the micro and mezzo levels, helping professionals can encourage our community members to recycle, avoid purchasing products that are dangerous to the environment, limit the use of fossil fuels, and buy from companies that reduce pollution. Many people become more motivated to participate in such new behaviors when they are part of a family or community that encourages such behaviors, so our small-group, institutional, and community-level work may be especially effective spaces to encourage behavioral change.

We can help educate our communities about the causes and effects of pollution. For example, there is growing evidence that air pollution inhibits airborne pollinators, including moths and many other insects, from locating flowers by their scent (Grandoni, 2024). Many people are unaware that air pollution can affect wildlife like this, and when we teach about these kinds of facts, more people become more concerned about these effects. Educational opportunities to teach pollution facts come up frequently in our work with community members.

At the macro level, many people may not be aware about how their own local politicians may be willing to oppose efforts to regulate major polluters and other common sources of air pollution. For example, although the federal government is pushing for stronger pollution standards to protect people from the effects of air pollution in the cities with the dirtiest air, some states are unfortunately pushing back against those standards (Siegler, 2024).

We can also help educate people that there are many forms of greenhouse gas (GHG) emissions, and some of the most dangerous are also probably the least well-known. Many people, for example, may not be aware of the dangers of human-caused methane gas emissions. *The Guardian* (Carrington, 2024) recently reported:

> Landfills emit methane when organic waste such as food scraps, wood, card, paper and garden waste decompose in the absence of oxygen. Methane, also called natural gas, traps 86 times more heat in the atmosphere than carbon dioxide over 20 years, making it a critical target for climate action. Scientists have said emissions from unmanaged landfills could double by 2050 as urban populations grow, blowing the chance of avoiding climate catastrophe. A total of 1,256 methane super-emitter events occurred between January 2019 and June 2023, according to the new data. ... Landfill emissions can be reduced by creating less organic waste in the first place, diverting it away from landfill, or at least capturing some of the methane that is being released from the landfills. Action to stem methane leaks slows global heating faster than almost any other measure and is often low-cost, with some measures even paying for themselves when the captured gas is sold as fuel. Methane emissions have accelerated since 2007 and cause a third of the global heating driving the climate crisis today. (pp. 1–2)

Water Pollution

Water pollution also is harmful to our biomes, but unfortunately, some politicians and influencers still oppose regulations that can help protect our water. According to *National Geographic* (Nuncz, 2020), U.S. freshwater pollution has been partly reduced by legislation that regulates the largest polluters. Because of these welcome controls, "nonpoint source pollution" is now our major challenge, which includes substances washed by rain and snowmelt over the ground into our water. This pollution often carries such contaminants as herbicides, pesticides, and fertilizers from neighborhoods, factories, and farms as well as livestock bacteria from ranches, industrial chemicals, and pet waste from our homes, sidewalks, streets, and parks.

Scientists are increasingly concerned with perfluoroalkyl and polyfluoroalkyl substances (PFAS) that are found in the products we buy that are designed to protect against heat, moisture, and stains. Many PFAS are called *forever chemicals* because of their persistence in the environment. All forms of water pollution can affect wildlife health, including the loss of species diversity and species extinction.

We are also becoming collectively more aware of the extent and dangers of plastic pollution. Much of the plastic that is released into in our rivers and lakes eventually reaches our oceans. Plastics typically deteriorate into tiny microplastics that are now increasingly found in our food, water, oceans, and bodies of animals. Plastic waste has been detected in human lungs, placental tissues, breast milk, and blood (Pinto-Rodrigues, 2023). Research is happening now to determine the health hazards associated with these microplastics in the bodies of many living things. Plastic waster also can be associated with other unforeseen negative consequences. For example, plastic trash concentrations can now act like life rafts for invasive species, which can float across our oceans and infect new locations (Stelle, 2023).

At the micro and mezzo levels of practice, we can encourage community members to change harmful habits that increase water pollution. To help reduce such pollution, we can encourage community members to dispose of toxic household products such as oil and paint safely. We can also use phosphate-free detergents and wash cars at commercial car washes that properly dispose of wastewater. Additionally, green roofs and rain gardens can help restore some of the natural filtering that forests and plants usually provide (Nunez, 2020). When people engage in such activities, they not only can help alleviate pollution but also often get a feeling of at least mild satisfaction that they are doing something positive about the climate crisis.

We can also educate people about the association between extreme weather events and water pollution. In a recent study, scientists looked at 965 case studies across the world and concluded that extreme weather harmed water quality on every habitable continent. During 51% of storms and flooding, significant fertilizer runoff was washed into rivers and streams. And during 68% of droughts and heat waves, the lower water volume resulted in greater concentration of pollution, including wastewater pharmaceuticals and found in wastewater. Overheated streams and oceans had lowered oxygen levels, endangering freshwater and marine life (*Yale Environment 360*, 2023).

At the macro level, parents and other community members can be encouraged and even organized to ask school boards to offer education to all children in grade, middle, and high school settings. High schools can also require dialogue groups in addition to (or to replace) the debate teams that we see in most high schools today. During such dialogues, students can learn dialogue skills and find ways to work with other community members to create new green policies at the local, state, and national levels. I have participated in these projects with high school students, in which they learn how to facilitate dialogue and then invite other students or community members to participate.

Soil Pollution

Soil pollution is largely human-caused, is often detrimental to plant growth, and disrupts food chains and our local and global biomes. Sources of soil pollution include liquid and solid contaminants intentionally or accidently released into the environment by industry, such as arsenic fluoride and sulfur dioxide. Industrial agricultural interests also release fertilizers, heavy metals, and sewage into our soils, and domestic and urban runoffs add contaminants as well. Deforestation, often a result of logging, typically increases soil erosion, which often reduces the amount of vegetation in its wake. Soil contamination reduces crops yields and quality and can enter biome ecosystems though the food chain. Soil contamination also can release volatile substances into the air and water, often through acid rain or algal blooms. Such spread of pollution can contribute to wildlife disease, loss of wildlife diversity, and species extinction (Clarke, 2022).

From 1960 to 2010, the amount the average person in the United States threw away tripled. Solid waste contributes to 5% of global GHG emissions, more than shipping or aviation. Most of the waste from high-income countries ends up in landfills, often in other countries (Franklin-Wallis, 2023). Such waste can contribute to soil, air, and water pollution.

The U.S. Environmental Protection Agency (EPA; Clarke, 2022) supports sustainable farming practices, including nutrient management, field buffers, tillage strategies, sustainable field drainage, regular ground cover, and livestock access to streams. The Pollution Prevention Act of 1990 helps reduce or eliminate pollution at the source. Government experts also can assist managers to reduce soil pollution in mines by implementing new waste disposal strategies, landscape restoration, and topsoil conservation. Wastewater treatment and other urban planning strategies can also reduce the urban sources of soil pollution, including the sewage we create. At the macro and eco levels, helping professionals can organize community members to support legislation and local policies that pay for such improvements as well as for enforcement of any similar legislation that may already exist.

At the micro and mezzo levels, much can be done by helping professionals to teach community members to reduce waste at home and their work locations. We can teach them how to work with local businesses, famers, and ranchers to reduce local soil pollution as well.

Light and Noise Pollution

Two-thirds of the world's human population live in light-polluted areas. Every year the proportion of the planet illuminated by artificial light grows by 1%, and the brightness of the illuminated area grows by 2%. Such artificial light can have profound negative impact on wildlife. For example, light is known to interfere with night-migrating birds, the tenuous march of sea turtle hatchlings to the sea, and the life cycles of many insects (Yong, 2022). Since there may be people who dismiss the importance of insects, we can remind them that our insects are actually a vital part of the ecology of most healthy biomes and that insects not only serve as food for many other creatures but also assist with such essential tasks as pollination and removal of dead organic material.

According to *National Geographic* (n.d.):

> Noise pollution can cause health problems for people and wildlife, both on land and in the sea. From traffic noise to rock concerts, loud or inescapable sounds can cause hearing loss, stress, and high blood pressure. Noise from ships and human activities in the ocean is harmful to whales and dolphins that depend on echolocation to survive. (p. 1)

At the micro and mezzo levels, we can educate community members about these kind of facts and encourage them to work together to reduce local light and noise pollution. I find that after I introduce my city folx to a night sky in a location without noise and light pollution, most are immediately more appreciative of the importance of a "natural" night sky.

We know that the noises of civilization disturb the migrations of many flying animals and the life cycles of sea mammals, fish, and other sea creatures. Such disturbances lead to the loss of species diversity. Although many animals can adapt to sound disturbances, many cannot (Yong, 2022).

At the macro and bio levels, the American Climate Corps, created by President Biden in 2023, gives people an opportunity to volunteer their time to make meaningful contributions to addressing the climate crisis. Their tasks might include forest maintenance and green construction in poor urban neighborhoods. A total of 20,000 young people will be hired to join the corps, and hopefully, they will go on to work in our future green economy (Siegler, 2023).

Marine Habitat and Land Habitat Destruction

Habitat destruction is associated many of the same factors associated with global warming, such as human-caused GHG emissions. According to the *National Geographic* (2019):

> Habitat destruction is one of the biggest threats facing plants and animal species throughout the world. The loss of habitat has far-reaching impacts on the planet's ability to sustain life, but even with the challenges, there is hope for the future. ... Habitat destruction, defined as the elimination or alteration of the conditions necessary for animals and plants to survive, not only impacts individual species but the health of the global ecosystem. Habitat loss is primarily, though not always, human-caused. (p. 1)

Although we are planting more trees and crops than ever, particularly in China and India, the amount of land and sea that supports the diverse plant life necessary for diverse animal populations is decreasing (Jones, 2024). Most of our marine wildlife live in coastal areas, streams, lakes, and oceans, and the majority of these animals breed in coastal estuaries. Unfortunately, water pollution, silt, and other dangerous materials are destroying these estuaries. In addition, about four-fifths of our global species live in forests. Sadly, we have lost about half of the trees on Earth since the start of civilization, and about 15 billion trees are cut down annually. Scientists have determined that if we protected 50% of the land and ocean around the world, our wild plant and animal life would be able to survive. Unfortunately, only 15% of the land and 7% of the ocean is currently protected (*National Geographic*, 2019).

We can help educate our communities about these kinds of dangers and encourage our students, patients, and clients to participate in organizations that seek to protect our essential biomes. For example, the Campaign for Nature (https://www.campaignfornature.org) calls on world leaders to take action in helping protect 30% of the Earth's land and ocean by 2030, on the way to a goal of keeping 50% of the planet in a natural state by 2050.

At the macro and bio levels, assisted migration is a solution that has been proposed as a remedy for warming climates and related loss of habitat. However, in the Pacific Northwest, controversy has developed regarding what kind of migration should be implemented with forests. There, a divide has emerged between groups advocating for assisted migration that would help struggling native trees and one that could instead see native species replaced on the landscape by trees from the South, including coast redwoods and giant sequoias.

According to Michael Case, a forest ecologist at the Arlington Nature Conservancy, there is a significant difference between assisted population migration and assisted species migration. Mr. Case supervises an assisted population migration program at the Ellsworth Creek Preserve in western Washington. Assisted population migration consists of moving the seeds (and therefore genes) of a native species within its current range. In contrast, assisted species migration involves moving a species a good distance outside its current range. (Gilles & the Associated Press, 2023).

Loss of Species Diversity and Species Extinctions

Scientists do not know the exact number of species of living beings currently exist on Earth. However, we do know that we are losing species at a rate between 1,000 and 10,000 times higher than the natural rate of extinction, which amounts to an estimated 200–100,000 annual extinction rate (WWF, 2023).

According to Paul Ehrlich (1988, Chapter 2):

> The primary cause of the decay of organic diversity is not direct human exploitation or malevolence, but the habitat destruction that inevitably results from the expansion of human populations

> and human activities. Many of the less cuddly, less spectacular organisms that *Homo sapiens* is wiping out are more important to the human future than are most of the publicized endangered species. ... By the time an organism is recognized as endangered, it is often too late to save it. ... Extrapolation of current trends in the reduction of diversity implies a denouement for civilization within the next 100 years comparable to a nuclear winter. ... Arresting the loss of diversity will be extremely difficult. The traditional "just set aside a preserve" approach is almost certain to be inadequate because of factors such as runaway human population growth, acid rains, and climate change induced by human beings. A quasi-religious transformation leading to the appreciation of diversity for its own sake, apart from the obvious direct benefits to humanity, may be required to save other organisms and ourselves.

I sometimes use the phrase *climate solutions* when discussing the climate crisis with individuals and groups . On NPR's *Morning Edition* in 2023, Julia Simon explained:

> Broadly speaking, climate solutions are things that reduce greenhouse gases—like solar and wind energy combined with batteries. Energy efficiency. Land use is key, too, like reducing deforestation. Individuals can play a role also—for example, eating less meat. ... But we have to remind folks that solutions are not all on individuals. A lot of solutions come down to companies and governments. For example, last year President Biden signed the Inflation Reduction Act—the most significant piece of climate policy in U.S. history.

One climate solution proposed in the literature is to create wildlife corridors that allow migrating and other species to safely travel from one biome area to another. Although conservationists see these corridors as a solution to species decline and extinction, other interests have a different view. The deaths of thousands of Wyoming pronghorn in winter 2022 increased interest in conservation leases. However, where wildlife advocates see hope, energy and ranching interests fear an attack on the "Western way of life" (Goldstein, 2023).

Often, Indigenous people lead the way in protecting the ecosystems in lands on which they once lived and called home. For example, the Yurok and other tribes help guard the Klamath River (Krol, 2023). Helping professionals can help link various groups who are all interested in such conservation groups. We can, for example, sponsor community dialogues in which members of many groups are invited to develop relationships and find common concerns and climate solutions.

We might want to believe that such legislation as the Endangered Species Act (ESA) automatically protects threatened species. However, there are obstacles in the way. Although the ESA requires that we save every species, there is a lack of funding and other resources necessary to do this (Kunzig, 2023). We can help teach our communities that such important legislation requires ongoing vigilance from citizens in order for it to work as it was intended. Citizens can encourage their local and national representatives to continue to support existing policies that were written to help protect biomes. For example, recent recognition of the 50-year birthday of the ESA included public celebrations that the Great Smoky Mountains National Park's wonderful biological diversity has been protected by the act for a half-century (Figart, 2023).

I have heard many students report benefit from participating in everyday household activities that help reduce waste and protect our biomes. For example, according to the EPA (2023):

> Composting is a process by which organic matter, such as leaves and food scraps, decomposes into soil. It's a great way to recycle scraps from your yard and kitchen while also enriching the soil in your garden, improving water retention, and protecting against erosion. Although it may sound complicated, composting is very simple and makes for a fun, rewarding hobby.

In his beautifully written book *Living Planet: The Web of Life on Earth,* David Attenborough (2021) names three international organizations that strive to protect species diversity: the International Union for the Conservation of Nature, the United Nations Environment Programme, and the WWF. He shares the following three principles on which these organizations have agreed regarding the responsible management of world wildlife:

1. We will not exploit natural populations of plants and animals to the extent that they are unable to renew themselves.
2. We will not change the face of the Earth to the extent that we damage the basic processes that sustain life, including the healthy systems in our atmosphere, seas, climate, and forests.
3. We will not damage the diversity of life, since we have no moral right to do so.

Vacations: Getting Away From It All

Many of us love to "get away" and go on vacation. Our word *vacation* is derived from the Latin *vacationem* which referred to leisure and freedom. Although most of us desire freedom and do not mind our occasional leisure, vacations can look vastly different for groups with different incomes and resources. In addition, in the United States, we often get very different messages about vacations from employers, politicians, and the advertising industry.

On the one hand, we constantly see advertisements about people enjoying vacations in beautiful biomes. For example, we may be shown happy people on cruise ships, being entertained and eating gourmet meals, while floating off sandy tropical beaches. Many commercials for new and expensive trucks and sport utility vehicles also show people smiling as they speed while on or off road in some pristine and scenic wild area.

On the other hand, the United States is the only developed country on Earth that does not require that workers receive paid time off. Indeed, workers in the United States generally receive much less and take less time off than workers in most other wealthy nations. Scientists have found, however, that vacations contribute to overall worker well-being, increase workers' appreciation of human diversity, and even enhance workers' productivity. There is also evidence that many U.S. workers feel guilty about taking time off and may be fearful that a vacation may negatively affect their job security (Taylor, 2019).

So, although may people of us in the United States understandably want to take vacation trips into the biomes that still survive, and although vacations are good for ourselves and our employers, we also seem to be reluctant as a society to take much vacation time.

In our desire to get away from it all, we seem to tend to emphasize the relationships we have with often distant biomes and deemphasize our relationships with the biomes in which we actually live. As Cronon (1996) aptly explains:

> If by definition wilderness leaves no place for human beings, save perhaps as contemplative sojourners enjoying their leisurely reverie in God's natural cathedral—then also by definition it

> can offer no solution to the environmental and other problems that confront us. ... Worse: to the extent that we live in an urban industrial civilization but at the same time pretend to ourselves that our real home is in the wilderness, to just that extent that we give ourselves permission to evade responsibility for the lives we actually lead. We inhabit civilization while holding some part of ourselves—what we imagine to be the most precious part—aloof from its entanglements. We work our nine-to-five jobs in its institutions, we eat its food, we drive its cars (not least to reach the wilderness), we benefit from the intricate and all too invisible networks with which it shelters us, all the while pretending that these things are not an essential part of who we are. By imagining that our true home is in the wilderness, we forgive ourselves the homes we actually inhabit.

Thus, our desire to get away from it all may often be associated with a desire to avoid taking responsibility for the climate crisis and especially to avoid responsibility for what is happening in the ecosystems in which we actually live for most of the year.

How can we helping professionals address these issues? At the macro level, we want to seek a middle-ground position that enhances both the well-being of workers as well as that of the ecosystems that keep us all alive. We want to support access to regular leave for all of our workers, including for those who belong to minoritized populations and receive very little to no nonwage compensation. We know that U.S. workers can benefit in many ways from having regular time off and that our workers have relatively few protections to guarantee that we all can receive such benefits. However, we also want to protect the ecosystems that support all life on our planet, including both our surviving wilderness biomes, where many of us vacation, as well as the "human biomes" in which most of us now live.

There are social and climate justice issues involved in our vacations today. At the macro level, we want to support policies that provide equitable access for all human populations, to our wild biomes, and to biomes in local urban areas.

At a micro level, we help people become conscious of any desire they may have to get away from it all. We have compassion for those that want to escape, because the times we live in are difficult indeed. It is painful to see what we humans have done to the Earth and especially painful to live in a region that is especially damaged by pollutants, ecosystem destruction, and intense construction. It is OK to want to have vacations (or, as the British call it, *holidays*).

It is also OK to take responsibility for the protection of both the biomes that we visit on vacations and the human biomes in which we live. Helping professionals can help educate community members about the vulnerabilities of our public lands and waters as well as the impact that humans have when we visit those special areas. We can educate everyone about the responsibility we all have to care for and protect the biomes we use for vacations and adventures.

For example, in the deserts of our Southwest, many people are not aware of the importance and fragility of biological soil crusts. These soil crusts are key microcommunities within desert biomes that are vital to the wellbeing of the entire biome.

> Biocrusts are complex communities composed of cyanobacteria, green algae, lichens, fungi, and mosses that form a living cover on many ground surfaces. They grow in most semiarid and arid ecosystems on this planet, from hot deserts to polar regions. These living crusts are the topsoil of the desert. ... They stabilize the soil, increase the soil's water absorption, aid in nutrient availability (particularly for vascular plants), and can enhance seedling establishment. (Williams, 2020, p. 16)

Scientists have found that when we walk or bike over such soil, its functions are restricted, and the recovery process may take from 50 to 250 years. A teacher could take children on field trips to examine firsthand the soil in nature and learn about its functions and vulnerabilities. The impressions of such an experience could last a lifetime.

Even the most remote areas of the planet need protection. Another striking example of the impact of humans on our most impressive biomes comes from the Himalayas. At the time of his first solo climb of the world's tallest peak, Reinhold Messner (1982) reported that Chomolungma (known in the West as *Mount Everest*) and its surrounding areas were already littered with garbage and waste from the many expeditions who visited the highest peak in the world.

Do we really want to leave for future generation these mountains of garbage over some of the most awesome mountain scenery on our planet? Similarly, do we want to create such monuments of destruction as polluted air, water, and soil and devastated ecosystems in so many of our urban landscapes?

QUESTIONS FOR REFLECTION

1. How has global warming affected the local biome in which you live? What impacts are you aware of on local wildlife?
2. How has pollution affected the local biome in which you live? What kind of psychological reactions do you notice you have when you look at the impact of local pollution?
3. What kinds of habitat destruction do you observe in the local biome in which you live? Is there any sense of loss that you feel about these changes? Please explain.
4. Do you know what kinds of wildlife existed hundreds or even thousands of years ago on the land and in the waters near where you now live? If you need to investigate this further, please find out more about your own local ecological past.
5. Do you know which losses of diversity and extinctions have occurred in the biome in which you live now? How do these losses affect you today? Please explain.
6. After reading the last section about vacations, what was your first emotional reaction? Do you see yourself in the words of Cronon (1996)? Please explain.

REFERENCES

Accius, J. (2024). The hidden climate crisis: Unequitable impact. *Fast Company*, February 9, 2024. https://www.fastcompany.com/91025411/the-hidden-climate-crisis-unequitable-impact

Attenborough, D. (2021). *Living planet: The web of life on Earth.* William Collins.

Caron, C. (2024). Therapists trade the couch for the great outdoors. *The New York Times*, February 5, 2024. https://www.nytimes.com/2024/02/05/well/mind/outdoor-therapy-depression-anxiety.html?bgrp=a&smid=em-share

Carrington, D. (2024). Revealed: The 1,200 big methane leaks from waste dumps trashing the planet. *The Guardian*, February 12, 2024. https://www.theguardian.com/environment/2024/feb/12/revealed-the-1200-big-methane-leaks-from-waste-dumps-trashing-the-planet?CMP=oth_b-aplnews_d-1

Clarke, R. (2022). *What is soil pollution? Environmental impacts and mitigation.* Tree Hugger, January 31, 2022. https://www.treehugger.com/what-is-soil-pollution-5194122

Cronon, W. (1996). The trouble with wilderness, or getting back to the wrong nature. In W. Cronon (Ed.), *Uncommon ground: Rethinking the human place in nature* (pp. 80–81). W. W. Norton.

Ehrlich, P. (1988). Chapter 2: The loss of biodiversity: Causes and consequences. In E. O. Wilson & F. M. Peter (Eds.), *Biodiversity.* National Academies Press. https://www.ncbi.nlm.nih.gov/books/NBK219310/

Environmental Protection Agency. (2023). *How to start composting at home.* https://www.healthline.com/nutrition/composting-beginners-guide

Figart, F. (2023). *At 50, Endangered Species Act continues to protect life in the park.* Smokies Life, December 21, 2023. https://smokieslife.org/2023/12/21/endangered-species-act-turns-50/

Franklin-Wallis, O. (2023). *Wasteland: The dirty truth about what we throw away, where it goes, and why it matters.* Simon & Schuster.

Gilles, N., & the Associated Press. (2023). Climate change is forcing Pacific Northwest trees to move—but scientists can't agree on what form of 'assisted migration' is best. *Fortune*, December 28, 2023. https://fortune.com/2023/12/28/climate-change-trees-migration-pacific-northwest-new-habitats/

Goldstein, A. (2023). A BLM proposal to protect wildlife corridors could restore the West's 'veins and arteries.' *Inside Climate News*, December 23, 2023. https://insideclimatenews.org/news/23122023/blm-protect-wildlife-corridors-the-west/

Grandoni, D. (2024). How air pollution prevents pollinators from finding their flowers. *The Washington Post*, February 8, 2024. https://www.washingtonpost.com/climate-environment/2024/02/08/air-pollution-pollinator-moths/

Jones, B. (2024). The Earth is literally getting greener. Here's why that worries scientists. *Vox*, February 7, 2024. https://www.vox.com/down-to-earth/2024/2/7/24057308/earth-global-greening-climate-change-carbon

Krol, D. (2023). Tribes guard the Klamath River's fish, water and lands as restoration begins at last. *USA Today*, December 29, 2023. https://www.usatoday.com/story/news/nation/2023/12/29/tribes-protect-water-land-lower-klamath-river/71938175007/

Kunzig, R. (2023). Can we save every species from extinction? *Scientific American.* https://www.scientificamerican.com/article/can-we-save-every-species-from-extinction/

Mann, B. (2023). *A naturalist finds hope despite climate change in an era he calls "the end of Eden."* National Public Radio, December 23, 2023. https://www.npr.org/2023/12/23/1214002178/naturalist-looks-for- hope-despite-climate-change

Mendiluce, M. (2024). Euroviews: Not all carbon credits are created equal. *euronews*, February 9, 2024. https://www.euronews.com/green/2024/02/09/not-all-carbon-credits-are-created-equal

Messner, R. (1982). *The crystal horizon: Everest: The first solo ascent.* The Mountaineers.

Milman, O. (2024). The world is reducing its reliance on fossil fuels—except for in three key sectors. *The Guardian*, February 9, 2024. https://www.theguardian.com/environment/2024/feb/09/biggest-fossil-fuel-emissions-shipping-plane-manufacturing?CMP=oth_b-aplnews_d-1

Mulvaney, K. (2024). 2023 was our hottest year yet. What other milestones did we reach? *National Geographic*, January 9, 2024. https://www.nationalgeographic.com/environment/article/signs-of-alarm-hope-for-the-environment-year-review

National Geographic. (n.d.). *Noise pollution.* Education. https://education.nationalgeographic.org/resource/noise-pollution/

National Geographic. (2019). *The global impacts of habitat destruction.* National Geographic Newsroom, September 25, 2019. https://blog.nationalgeographic.org/2019/09/25/the-global-impacts-of-habitat-destruction/

Nunez, C. (2020). Water pollution is a rising global crisis. Here's what you need to know. *National Geographic*, January 24, 2020. https://www.nationalgeographic.com/environment/article/freshwater-pollution

Pengue, M. (2022). *How does air pollution affect animals? The irreparable damage to wildlife.* Seed Scientific, August 15, 2022. https://seedscientific.com/environment/how-does-air-pollution-affect-animals/

Pinto-Rodrigues, A. (2023). *Microplastics are in our bodies. Here's why we don't know the health risks.* Science News, March 24, 2023.

Potts, M., Burton, C., Yang, T., Radcliffe, M., & Rakich, N. (2024). *Americans believe in climate change, but do they believe in love?* ABC News, February 9, 2024. https://abcnews.go.com/538/americans-climate-change-love/story?id=107027932

Siegler, K. (2023). *Biden creates the American Climate Corps, 90 years after FDR put 3 million to work in national parks.* Living on Earth, September 30, 2023. https://insideclimatenews.org/news/30092023/biden-createas-the-american-climate-corps/

Siegler, K. (2024). Utah is pushing back against ever-tightening EPA air pollution standards. National Public Radio, *All Things Considered*, February 9, 2024. https://www.npr.org/2024/02/09/1230040881/utah-is-pushing-back-against-ever-tightening-epa-air-pollution-standards

Simon, J. (2023). Climate solutions are necessary. So we're dedicating a week to highlight them. National Public Radio, *Morning Edition*, October 2, 2023. https://www.npr.org/2023/10/02/1197590139/climate-change-solutions

Stelle, L. (2023). *Scientists discover alarming new effect of plastic pollution in the ocean: '[Things are] rapidly changing.'* The Cool Down, December 29, 2023. https://www.thecooldown.com/outdoors/great-pacific-garbage-patch-plastic-invasive-species/

Stone, E. (2020). Geological history of the boundary waters. *Northern Wilds*, April 1, 2020. https://northernwilds.com/geologic-history-of-the-boundary-waters/

Taylor, T. (2019). *Why do Americans feel guilty about using vacation benefits?* The Balance, July 18, 2019. https://www.thebalancemoney.com/why-americans-feel-guilty-about-using-vacation-benefits-4156884

U.S. Geological Survey. (2023). *Climate impacts on plants and animals.* https://www.usgs.gov/science/science-explorer/climate/impacts-on-plants-and-animals

Williams, D. B. (2020). *A naturalist's guide to canyon country.* Falcon Guides.

World Wildlife Federation. (2018). *The Amazon, Miombo Woodlands in Southern Africa, and Southwest Australia will be among the most affected places in the world, according to comprehensive new paper and report commissioned by WWF.* March 14, 2018. https://www.wwf.org.uk/updates/climate-change-affects-50-plant-and-animal-species

World Wildlife Federation. (2023). *How many species are we losing?* https://wwf.panda.org/discover/our_focus/biodiversity/biodiversity/

Yale Environment 360. (2023). Climate change hurting water quality in rivers worldwide. September 12, 2023. https://e360.yale.edu/digest/climate-change-water-quality-rivers

Yoder, K. (2023). Climate wrapped: Top 10 terms that capture the climate in 2023. *Mother Jones*, December 29, 2023. https://www.motherjones.com/environment/2023/12/climate-wrapped-top-10-terms-that-capture-the-climate-in-2023/

Yong, E. (2022). *An immense world: How animal senses reveal the hidden realms around us.* Random House.

Credit

IMG 14.1-3

Practice With Anthromes

Human societies and their use of land have transformed ecology across this planet for thousands of years. As a result, the global patterns of life on Earth, the biomes, can no longer be understood without considering how humans have altered them. Anthromes, or anthropogenic biomes, characterize the globally significant ecological patterns created by sustained direct human interactions with ecosystems, including agriculture, urbanization, and other land uses. Anthromes now cover more than three quarters of Earth's ice-free land surface, including dense settlements, villages, croplands, rangelands, and seminatural lands; wildlands untransformed by agriculture and settlements cover the remaining area.

—ERLE C. ELLIS, IN *ANTHROMES* (2020)

I come here today with a message: As President, I have a responsibility to act with urgency and resolve when our nation faces clear and present danger. And that's what climate change is about. It is literally, not figuratively, a clear and present danger.

The health of our citizens and our communities is literally at stake.

—PRESIDENT JOE BIDEN (JULY 20, 2022)

The first summer after I moved to Salt Lake City, I did a solo backpack trip in the Uinta Mountains. They are the highest mountains in the state, and much of the region is designated as roadless wilderness. The biggest peaks range from elevations of 11,000 feet to the 13,528-foot summit of Kings Peak. Pre-Cambrian rock more than 600 million years old formed the core of the Uintas, which are also the only East-West–oriented major range in the contiguous United States.

Some of the valleys and basins are among the prettiest mountain country I have ever seen, with meadows and forests framed by ridges that tower several thousand feet above.

After a long bumpy drive after dark along an old gravel road and a cold night sleeping under the Milky Way, I started out early up the trail toward the top of the north slope. A day into the wilderness, as

I crossed over a pass, I felt a deep rumbling ahead. What could be causing such a sensation out here, so far away from the city? Maybe an earthquake?

Then I saw the cause of the disturbance: It was a large herd of sheep, crossing the meadow in a group of several hundred animals.

A little bummed that I wasn't as alone as I thought, I had thought the Uinta wilderness was untouched by grazing. When I returned home and looked it up online, I found out that domestic sheep grazing was still allowed in the High Uintas Wilderness, which meant that even this remote and seemingly pristine area could be considered an anthrome.

As we discussed in earlier chapters, biomes and anthromes can take many forms, and the term *anthrome* is now itself controversial. According to Erle Ellis (2020):

> Biomes are introduced as the fundamental units of the biosphere in nearly every introductory biology, Earth science, and environmental science textbook. While anthromes should not be thought of as a replacement for existing biome systems based on vegetation and climate, anthromes offer a more objective view of the terrestrial biosphere in its contemporary anthropogenic state. This new model of the biosphere moves us away from an outdated view of the world as "natural ecosystems with humans disturbing them" and towards a vision of "human systems with natural ecosystems entangled with them" (also known as social-ecological systems, anthroecosystems, or coupled human and natural systems).

We can identify six land-use levels of anthromes (Anthroecology Lab, 2023), which are dense settlements, villages, croplands, rangelands, seminatural lands, and wildlands (see Table 14.1). Examples of a dense settlement (e.g., Boston), cropland (e.g., Sacramento Valley), and wildland (e.g., Southwestern mountainside) are found in the images at the beginning of this chapter.

TABLE 14.1 Anthrome Levels

Anthrome Level	Global Percentage	Description
Dense settlements	1.7%	Urban and other nonagricultural dense settlements
Villages	7.1%	Densely populated agricultural settlements
Croplands	14.7%	Lands used mainly for annual crops
Rangelands	26.5%	Lands used for pasture and livestock grazing
Seminatural lands	24.3%	Inhabited lands with minor use for permanent agriculture and settlements
Wildlands	25.8%	Lands without human populations or substantial land use

Source: Anthroecology Lab, 2023.

Thus, practice with anthromes by helping professionals includes work with varying degrees of emphasis upon the well-being of humans, other living things, and various local ecosystems. For example, practice in an urban environment such as Boston might especially focus on such areas as human well-being in the dense population living in that environment as well as on the quality of water and ocean life off the local shores; local air; local streams, lakes, and parklands; and perhaps the political polarization that interferes with progress in building sustainable green power systems. In contrast, practice in the croplands of the Sacramento Valley, which has a much smaller human population density, might focus more on such areas as the impact of irrigation on underground water levels, use of runoff from annual mountain snowpack, effect of fertilizers and pesticides on local environment, and, perhaps, influence of new green energy technologies on local wildlife and humanlife.

Climate-sensitive practice, introduced in Section 2 of this text, may use any of the practice approaches described in this chapter.

In this chapter, as we did in Chapter 13, we will again organize content by the practice topics that we will address. These topics are listed in Table 14.2. They are all issues humans face in the more densely human-populated anthromes on Earth. Building on the climate facts we know from the most recent literature, we will focus on selected climate solutions in each of the sections.

TABLE 14.2 Practice Topics

Fostering faith-based activism
Developing activism wisdom
Mitigating polarization and outrage
Mitigating the impacts of weather and climate disasters
Mitigating our public health climate crisis
Shifting from war economies to peace economies
Making better climate solutions

Fostering Faith-Based Optimism

Throughout our text, we have examined how our climate crisis can be associated with human hope and despair. The daily news we watch about global warming, pollution, and biodiversity loss can be scary and demoralizing.

Data scientist Hanne Ritchie (2024) in her recent book *Not the End of the World: How We Can Be the First Generation to Build a Sustainable Planet* argues that there are many reasons for us to start believing that we can successfully address our climate crisis. She cites ways that we have and can make a difference in such areas as climate change, air pollution, food, deforestation, biodiversity loss, ocean plastics, and overfishing.

In her review of Ritchie's text, Bibi van der Zee (2024) points out that Ritchie does not address:

> the domestic and geopolitical barriers, together with the inbuilt biases and quirks of our brains, that combine to make environmental issues so difficult to address. We know now that humans don't like to give things up; we are afraid of losing what we already have and afraid of change. We are brilliant at inventing things and making great leaps of imagination, but we are terrible at looking

> into the future, and at understanding the risks attached to those inventions. This applies equally to world leaders and business executives. Of course the heads of huge fossil fuel companies and petro-states are going to be reluctant to give up the things that make them rich and keep them in power, something we saw play out again at COP28 in Dubai [the 2023 United Nations Climate Change Conference] . ... Vast numbers of us do want to work towards the beautiful sustainable society that Ritchie has in mind. But there are other groups, fueled by anger or fear or greed, that really do not.

Can we sustain our faith-based activism in the face of the domestic and geopolitical barriers and internal biases that challenge us every day? I believe that the answer is yes. Let's explore some approaches to our practice with anthromes that may be helpful in sustaining our own activism.

We have explored through this text the fact that on any given day, there is a mix of both discouraging and encouraging climate news. For example, in a *Scientific American* article, Andrea Thompson (2023) reported on how the biggest climate stories of 2023 were a mixed collection of both good and bad news. On the one hand, for example, July was the hottest month ever recorded on the planet, and frequent and severe heat waves and flooding disasters bankrupted more private insurers in states such as California, Florida, and Louisiana. On the other hand, the Biden administration used rulemaking, executive action, and international diplomacy to reduce U.S. greenhouse gas (GHG) emissions, increased environmental justice rules, and created an American Climate Corps. And on the international level, the United States and China agreed to expand renewable energy and carbon capture programs.

In addition, positive and heartening stories may not make the news as frequently as negative and scary stories that may seem to gather more attention. For example, Paige Vega, climate editor of *Vox*, has published a number of articles that:

> offer readers stories about community activism and agency, emerging science that provides hope in a mining-torn landscape, recovery and resilience, and a glimpse of the coastal communities standing up to El Niño's wrath. We aim to contextualize this moment and explore big themes such as the notions of refuge, the power of community activism, and the promise of climate reparations for nations in need. The climate crisis—and the decisions we make to address it—are already impacting countries around the world; here, we explain what's at stake through the lens of our international communities. (Vega, 2023)

We are also discovering that psychotherapists use a variety of strategies to help their clients deal more effectively with climate anxiety (Haupt, 2024). They may encourage clients to join dialogue groups that discuss climate crisis issues. The Climate Café Hub (https://www.climate.cafe/) provides contact information for both in-person and online groups. Therapists may also suggest that clients join their families or communities to take action. Clients are also asked to determine the amount of GHGs that they produce directly, perhaps by using the online tool found at https://www.nature.org/en-us/get-involved/how-to-help/carbon-footprint-calculator/. Clients may also find that they benefit from talking with others about their feelings and thoughts or expressing themselves through art. Finally, time outdoors also may benefit many clients.

There are many small ways in which each of us can make a difference and that enable us to practice *both* self-empowerment and responsibility to our communities. Our students, clients, and patients might, for example, be interested in changing their diet, which can be a relatively simple way for them to both improve their health as well as self-empower themselves to make a difference in their world. Italian researchers, for example, have recently found that the implementation of plant-based diets can both reduce the risk of cardiovascular disease and help reduce GHG emissions (Marino, 2024).

Developing Activism Wisdom

One way to bridge the differences that divide us is to develop what we could call *activism wisdom*. Activism wisdom involves using the best ethical approaches to fostering climate change that are most likely to work.

The *first* part of activism wisdom—and probably the most important—is to check our own ego. Perhaps this is difficult for many of us to do, precisely because we most of us, on all sides of any issue, believe that we are in the right. Although there is nothing wrong, of course, with believing that I am on the right side of an issue, with such conviction comes the increased danger that I might start to identify with my beliefs. Such identification is synonymous with the definition of *ego* that we used early in the text. In other words, we may fall victim to the strong temptation to give in to our ego, begin to feel superior to those who disagree with me, and, ultimately, even seek to attack them. Other people, of course, can sense when another person feels superior to them, and most do not like that attitude.

We start with self-reflection and meditation so that we can see ourselves the way we actually are, with reverence for what we find. We all can benefit from studying our own mind. I want to know what my motivations for being a climate activist are, for example. I want to become conscious of my own ego, my own strengths and limitations, and how love and connection can bring people together.

Second, activism wisdom means that we pay attention to what we say. Although she reports that 80% of the human world support doing "whatever it takes" to address climate change, Kate Yoder (2023a) also advises us to think about *how* we try to influence others. She writes:

> While people around the world are united in supporting government action on climate change, some of that support evaporated when it came to specific policies. They were most enthusiastic about clean energy instead of coal and subsidies for renewable energy companies, and least enthusiastic for phasing out fossil fuels and ending subsidies for polluters. Messages that used the words "mandate," "ban," or "phaseout" generated 9 percentage points less support, on average, than those that didn't. For example, only 54% were in favor of "banning" gas appliances in buildings, but 74% approved of requiring "better technologies" and "smart upgrades" in all new construction. That could be bad news for popular climate catch phrases like "keep it in the ground." In other words, we people do not like being told what to do.

Yoder (2023a) also quotes John Marshall, chief executive officer of the nonpartisan, nonprofit marketing firm Potential Energy Coalition:

> "The data is saying we need to lean in to the messages that get us the wins, as opposed to the messages that make us feel good about ourselves. Talking about upgrading appliances and heating and cooling systems and setting clean energy goals increased people's support for climate policies. The only kind of limitation people liked was reducing pollution. For that reason, it's important to stress that burning fossil fuels causes pollution that's overheating the planet."

Therefore, I want to become conscious of the world and also see it as it actually is, with a reverence for what exists now as well. That includes seeing human nature realistically. I especially seek to understand and have compassion for those who disagree with me. I want to understand the climate crisis and the role we humans have played in creating it.

Third, people want to enjoy their interactions with other people. Our activism wisdom reminds us to be civil with each other and enjoy ourselves and each other as we work together to make a difference in our local and global communities. Throughout our text, we have defined and applied *dialogue practice* to our climate practice. An essential part of dialogue is civility, and a healthy community—including an activist community—practices civility toward each other and other people as well. And, as we have discussed in the text, there is evidence of a loneliness epidemic in the United States today (Rodriguez, 2024). Many people are especially hungry for community and are most likely to participate in inclusive groups that are welcoming and supportive communities.

Fourth, I believe that the best way to change the world is often through relationships. We can influence people by treating them with loving kindness and forgiveness and modeling new behaviors in integrity. We talk about *loving kindness,* rather than about *love,* because *loving kindness* is both an attitude and a behavior. As Dr. Martin Luther King, Jr. (2012) said, "Returning hate for hate multiplies hate, adding deeper darkness to a night already devoid of stars. Darkness cannot drive out darkness; only light can do that. Hate cannot drive out hate; only love can do that."

I especially like an article written by Michael Coren (2023) for *The Washington Post* in which he reminds us that we can have a direct positive influence on the people we know in our lives:

> Your decision to buy that heat pump or induction stove might feel like it came after much deliberation and research. You might want to thank your friends and family. Your trusted inner circle is one of the most potent and overlooked weapons to stave off the worst of climate change. Our individual actions appear small, but they act as billboards for others looking for cues on what to do in their own lives. These social comparisons can add up.

As helping professionals, we can remind other people in our community about this uncomplicated truth and encourage them to model the changes they want to see in other people. We also want to make sure that *we* do not forget that people in our community pay attention to how we live our life and that living in integrity, walking our talk, may be the most powerful climate interventions we can make.

As an activist, I strive to act with integrity in my life, at home, at work, and in my community. I seek to model the behaviors that I want to see in others, including loving kindness, truth-telling, and service to other beings and to the world we all share. Behavior modeling can be a powerful way to influence others. The word *discipline* is related to *disciple,* which meant "follower" in Old English. People tend to follow the example of others whom they respect.

Finally, *fifth,* we practice dialogue rather than attack. We have discussed the elements and benefits of dialogue throughout most of the text. Again, the purpose of dialogue is to help people bridge the differences that currently divide them. The essential dialogue elements of listening for understanding and speaking with respect can help faith-based climate activists begin to heal and transform their own local communities.

Mitigating Polarization and Outrage

We addressed polarization in Chapter 7 and reexamine it here, as we look at our practice with anthromes. As we read in the van der Zee (2024) quote earlier, we cannot ignore these "domestic and geopolitical barriers" that seem so often to stand in the way of progress in addressing our climate crisis. These obstacles include the polarization and outrage that seem so endemic today across our world. What are the factors that contribute to them, and what can be done about it?

Unfortunately, the important work that we need to do to address the climate crisis is often delayed and hindered by special interests, whose activities contribute to the political polarization and public outrage that so many of us say they do not want. A recent survey found that many U.S. citizens report that they are "burned out on outrage" and that about two-thirds of us report often feeling exhausted when thinking about politics (Glueck, 2023).

According to U.S. Special Presidential Envoy for Climate John Kerry,

> people are not being told the truth about what the impacts are from making this transition [to net-zero GHG emissions]. They're being scared, purposely frightened by the demagoguery that is oblivious to the facts or distorting the facts. And in some cases outright lying is going on. ... We have politics now entering into this—fighting for delay and fighting for progress. They're procrastinating and they're part of the disinformation crowd that are willing to put the whole world at risk for whatever political motivations may be behind their choices here. (Harvey, 2024, pp. 1–2)

Kerry warned that the goal of special interests are to make people doubt our solid climate science, remove the United States from the Paris Agreement, and reverse the historic clean energy policies that the Biden administration has worked so hard to establish (Harvey, 2024, pp. 1–2).

In his book *Outrage Machine: How Tech Amplifies Discontent, Disrupts Democracies—and What We Can Do About It*, Tobias Rose-Stockwell (2023) argues that the biggest contribution communications technology has made to the political polarization that we live with today is the wide-spread acceptance and even glorification of public outrage. He states that "social media inflamed 'faction' or tribalism, and ... was now breaking society, civility, shared meaning, and democracy ... a machine designed to make you angry ... [also offering] meaningful gifts of connection, rights, and insights ... but to use it, we must play by the rules. The rules are currently rigged against us."

Our current communications technology is just the latest in a series of new technological advances that emerged over the centuries. Social media began as a wonderful experiment that promised to help connect us together. Instead, it now exploits our innate xenophobic tendencies and encourages us to express outrage rather than to reflect and exercise self-restraint. Humans seem to have evolved to especially notice and react to strong emotions, he writes, and these media platforms now reward outrage and penalize tolerance and allow politicians and other actors to deepen existing political divisions. He recommends that users limit screen time and, instead of fighting with strangers, focus on building relationships with them (Rose-Stockwell, 2023).

In *They Knew: How a Culture of Conspiracy Keeps America Complacent*, Sarah Kendzior (2022) argues that the U.S. population is now especially susceptible to fake conspiracies largely because we do not enforce accountability for real conspiracies. From this perspective, we sense correctly that there are real conspiracies and real corruption, seen, for example, in the ways that some large corporations seek to avoid accountability for their production or burning of fuels that produce dangerous GHGs. However, instead of pursuing such real issues, it can be easier for us to focus upon the conspiracies that we are fed in the "echo-chamber" media programs that many of us tend to watch faithfully. Kendzior argues that we all need to work at distinguishing the truth from fiction so that we can together foster increased accountability and social justice in our communities.

Political polarization may ease as politicians from all parties recognize how the climate crisis is associated with our economy, immigration, crime, and inequality (Young, 2023). However, we need to stay vigilant, because in every era, there will be politicians who focus on cultural differences and take extreme positions in their political campaigns rather than seek compromises through civil conversation that supports the well-being of everyone. There are signals, for example, that if he is reelected, a new former President Donald Trump administration will try to reverse the Biden administration's efforts to address the climate crisis, according to Republican strategy experts and Trump's own campaign website (Reuters, 2024). Therefore, we are likely to see such policies as more gas and oil drilling, another withdrawal from the Paris Agreement, and federal support for the proliferation of gas-powered vehicles (Reuters, 2024). As we have discussed, there is typically a middle ground that can be found in any conflict, and professional helpers can encourage community members to seek local cooperation and ask their political representatives to do the same on state and national levels.

Polarization and outrage exists not only between the left and right but also *within* political parties and environmental groups. For decades I have sat on many committees and other meetings that are devoted to championing a cause that I support. Usually most of the people in the room are similarly committed to the goal; however, they often disagree *unkindly* with each other about the means to it. Along these lines, Ritchie (2023) explains:

> One of the most effective ways to be a climate sceptic is to say nothing at all. Why expend the effort slapping down climate solutions when you can rely on feuding climate activists to tear each other's ideas apart? We tend to fight with those we are closest to. This is true of family. But it's also true of our peers, which for me, are those obsessed with trying to fix climate change. Step into the murky waters of Twitter and you'll often find activists spending more time going after one another than battling climate falsehoods. ... Sure, we all have our favourite solutions. But the reality is that we can't afford to be choosy. The answer to almost every climate dilemma is "We need both." (p. 1)

Ritchie makes an important point. We can all become attached to a particular climate issue or climate solution; however, often, when we look closely, we can recognize that many issues require scrutiny and many solutions may need to be tested. That does not mean that priorities have to be made when we face critical issues and time and funding are limited, but the climate crisis is associated with many complex factors, and we need to examine all the issues and solutions carefully. We may have to depend on *both* nuclear power and renewables, for example, during the coming decades. We may have to focus on *both* electric vehicles (EVs) and public transportation in our urban areas. And we need to work on establishing a climate change in the three practice areas we have investigated in this text—in human consciousness, humansystems, *and* ecosystems.

It may well be that legal action is necessary when other attempts to communicate have failed to stop the creation and spread of climate disinformation and perhaps also help reduce public outrage. National Public

Radio (NPR) recently reported that climate scientist Michael Mann was awarded more than $1 million in damages in a defamation case he brought a decade ago against conservative authors (Simon, 2024). Once again, we want to encourage conversation about the differences that exist in our communities; however, we ask that such conversation remain civil.

Helping professionals can assist community members in understanding the psychological roots of misinformation and helping reduce public outrage and polarization. Andy Norman is cofounder of the Mental Immunity Project, which helps protect people from misinformation. He has found that in these often difficult times, people can overlook inconvenient facts that do not support their beliefs. They also may enjoy believing that they know the real truth and that they can see past the illusions they think most people hold onto, despite the fact that the opposite is true (Yoder, 2023b). When we become more conscious about exploring what is going on inside ourselves—as we discussed in Section 1—we become less vulnerable to believing and spreading misinformation and disinformation. Certain sources of climate news strive to provide the public with accurate information about what is happening in that field. One such source of climate news can be found at Inside Climate News (https://insideclimatenews.org/), which is the website of a Pulitzer Prize–winning, nonpartisan newsletter devoted to exploring climate crisis stories. A quick view of the website on February 24, 2024, for example, found four fascinating stories, including news regarding how the Bad River Band of the Lake Superior Tribe of Chippewa Indians is fighting a pipeline, many people living near the East Palestine train derailment site say they're still sick, a "people's tribunal" in California asked the attorney general to investigate state and county pesticide regulators who may not be protecting farmworkers and their families, and Los Angeles had about 600 landslides in one week after recent coastal flooding.

Helping professionals can also point our students, patients, and clients toward reputable sources that regularly report breaking science news. For example, *Popular Mechanics* (Newcomb, 2024) recently reported that researchers have found that oxygen isotopes in various rock samples show that, going back to the Proterozoic era (spanning the time interval 2,500–538.8 million years ago), Earth was never hotter than 140 degrees Fahrenheit (F) and ocean temperatures have remained stable throughout that time. Their conclusion was that Earth experienced temperate conditions that supported the evolution and proliferation of life on our planet. These kinds of data help reinforce our understanding that the rapidly increasing and extreme heat we are experiencing today is largely human-caused. We are fortunate in many ways that we now have such easy access to information even though we are simultaneously challenged by the misuse of media technology through the spread of disinformation and misinformation.

Mitigating the Impacts of Weather and Climate Disasters

Many, if not most, of us have noticed that the frequency and extent of weather disasters (more short-term and local) and climate disasters (more short-term and widespread) seem to be increasing. These catastrophes may happen in any season and any location on Earth, although some regions and populations are more vulnerable, as we will discuss.

We may not be as aware of *tipping points,* which are self-sustaining climate changes that can result in serious and irreversible changes. These points can be activated by the high levels of GHGs currently being emitted. Many key parts of our biomes can be tipped, including ocean levels, ice sheets, ocean currents, and forests. Scientists are still uncertain about exactly what the thresholds are for these tipping points.

The Paris Agreement's goal of keeping global warming below 1.5 degrees Celsius (C) would reduce the chances of triggering multiple climate tipping points (McKay et al., 2022).

In winter 2023, NPR (Sommer, 2023) reported that the United States will suffer three major climate outcomes if global warming rises above 1.5 degrees C (2.7 F). First, the rate of climate change will increase, and the United States will heat up even faster than before. Second, our rainfall will become more intense as tropical storms and other severe rainfall events bring more flooding. Lastly, our extreme heat events will also worsen, and we will experience a greater number of hot days and a fewer number of cold ones.

Rising temperatures have a growing impact on where people move to and from. Over the next three decades, scientists predict that the most intense climate disasters will hit the East Coast, including Florida, the Carolinas, and the New England area, due to wind, flooding, and sea level rise, and the Southwest, including New Mexico, Utah, and Arizona, due to excessive heat, fires, and dwindling water supply. An estimated two-thirds of the U.S. population report that climate and weather extremes are factors that influence their decisions when planning possible future moves. The prevalence of these disasters have grown quickly and are increasingly common in at least 37 U.S. states (Feiger, 2023).

In fact, since 1980, the United States has suffered 376 climate and weather disasters with damages of at least $1 billion, exceeding a total cost to the country of $2.655 trillion. In 2023, for example, there were 28 weather and/or climate disaster events of at least $1 billion, including drought, floods, severe storms, tropical cyclones, wildfires, and winter storms. These events caused 492 human deaths (National Centers for Environmental Information, 2024). Similar kinds of weather and climate disasters have occurred on a worldwide level as well (Pulver, 2023). A study that was released as I was finishing this text indicated that climate change may cost the planet $38 trillion a year by 2049, which corresponds to a loss of 19% of income per capita globally within the next 26 years (Freedman, 2024).

A goal of limiting global warming since preindustrial levels to 2 degrees C was part of a 2015 agreement signed by 196 signatories to the Paris Agreement. Many scientists preferred a goal of 1.5 degrees C, warning that any temperature increase increases the chances that tipping points will be reached. Some kind of temperature goal was deemed necessary; therefore, the 2 degrees C goal was as much political as scientific (Mulvaney, 2023).

A summary review of recent studies (Carnell, 2024) has determined that hurricane wind speeds keep rising to such an extent that a new scale of wind speeds and related damages may become necessary to include these larger figures. In addition to 2023 being the hottest year on record, scientists have also discovered that Earth may have already experienced the warming they predicted to happen decades from now in 2100. More than 4 million people are now believed to have died from the climate crisis since 2000. Finally, as winter weather extremes become the norm, our snowpack and freshwater have reached new record low levels.

With these complex challenges facing us, it becomes increasingly important to find the most effective ways to explore and respond to our climate crisis. Participants in the recent Climate Now debate on climate action (Hughes, 2023) noted this urgency as they offered suggestions for climate change strategies. Panelist Lucy Hubble-Rose suggested:

> It's more important to solve an economic problem in a way that is also helpful for climate change rather than taking a climate change action that is also helpful for the economy. ... Similarly, within governments and councils, climate action needs to be integrated in other policymaking. ... We know that building your own individual sense of agency is really important, so we want to tell people

great stories of the sorts of action they can take so they can get past that feeling of being stuck. ... It seems counterintuitive, but we find that actions drive beliefs rather than the other way round.

These comments remind me of the advice I received from an Iowan community activist back in 1972, who told me, “It is often faster to act our way into new ways of thinking than to try to think our way into new ways of acting.” I have seen the wisdom in his words when working on the micro, mezzo, macro, and eco levels of practice.

Those panelists also addressed the importance of designing early warning systems to help mitigate weather and climate emergencies, including evacuation planning and hospital preparations for dealing with a heavy influx of patients during floods and heatwaves. They discussed how to cool urban areas more efficiently, including conducting thermal building renovation, using heat-resistant colors in roof paint, promoting swimming, and canceling summer festivals during extreme heat waves.

Youmshajekian and Fragapane (2024) have developed a flow chart that can be very useful in teaching community members about the multiple levels of outcomes that can come from climate disasters in the anthromes in which they live (see Table 14.3). In this article, the authors present a graphic display for their information, which the reader might want to review.

TABLE 14.3 Climate Hazard

Climate Hazards	Secondary Hazards	Short-Term Outcomes	Long-Term Outcomes
Extended heat wave	Ground-level ozone	Dehydration Heat stroke	Mortality Mental health
Drought	Earlier snowmelt runoff Water shortage Food prices	Food insecurity Communicable disease	Mortality Mental health Malnutrition
Wildfire	Air pollution Evacuation Property loss	Displacement Mortality Injury	Mortality Severe burns Other physical health effects
Hurricane and storm surge	Water contamination Exposure to flood water	Stress Gastrointestinal illness	Mortality Mental health Other physical health effects
Wind	Evacuation Resident destruction	Cleanup reconstruction Allergies	Mortality Mental health
Intense precipitation and flooding	Power outage Water damage	Cold Heat Emergency medical needs	Mortality Mental health Other physical health effects

Source: Youmshajekian & Fragapane, 2024.

Ultimately, probably the best way to mitigate weather and climate disasters is to reduce GHG emissions more quickly and profoundly. However, as helping professionals, we realize that, given the current state of the world, we are wise to include both prevention and intervention strategies when making our mitigation plans.

Mitigating Our Public Health Climate Crisis

The climate crisis is also a public health crisis. As we have seen, on a global level, scientists are recording more frequent and intense incidents of such weather extremes as heat waves, droughts, cold snaps, dangerous storms, and floods, all of which create public health hazards. In California, for example, following recent cycles of draught and flooding, a fungal disease known as *coccidioidomycosis* or *valley fever* has thrived. These fungi live in wet soil, and in dry conditions, their spores later blow into the air and into human respiratory systems, where they cause pneumonialike symptoms of cough and fever. There were more than 9,280 new cases of valley fever in 2023. Similarly, when Tropical Storm Hilary crossed through California, it brought a 560% increase in diarrheal illness cases in the areas most affected (Price, 2024).

As we reviewed in Chapter 8, about climate justice, vulnerable people are likely to be especially exposed to health risks when their homes are destroyed; energy, food, and water shortages occur; and evacuation and medical care become necessary (Sattler, 2024). Part of the mitigation of the climate public health crisis must be a focus on climate justice. Minoritized people pay both a material and a psychological toll in a world where a "polluter elite" of fewer than 1% of the global population produce as much carbon emissions as 5 billion poorer people and the richest 10% are responsible for about 50% of all emissions (Cohen, 2023).

More than 3 million people in the United States have already become climate migrants as they retreat from such worsening conditions as extreme heat, drought, and flooding (Kaufman, 2023). Leaders representing the most vulnerable regions across the world often express concern that recent global climate deals have largely served to appease the fossil fuel industry and have not gone far enough to protect them from climate crisis–related events such as climate migration (Friedman, 2023). In these vulnerable regions, when migration becomes necessary, members of already minoritized populations are especially affected because they typically have relatively few resources and options.

Across our planet, aging populations make up one of the groups most vulnerable to climate and weather extremes. Aging people tend to be susceptible to heat and cold and are often less able to escape such events as major snowfall, flooding, and fire (Arigoni, 2023). Table 14.4 lists basic strategies for helping build age-friendly resilience for the aging population that can also be adopted by other populations at risk.

TABLE 14.4 Strategies for Age-Friendly Resilience

Target Systems	Examples of Key Target Elements
Energy	Affordability Reliability
Housing	Affordability Security from weather hazards

Transportation	Affordability Reliability Mobility options Accessibility
Social infrastructure	Accessible media communications Strong social networks Offering home health care aides disaster training
Health care	Affordability Climate education for workers and patients Telehealth and mobile clinics
Emergency management	Hazard mitigation plans' inclusion of vulnerable populations Multipurpose resiliency centers

Source: Arigoni, 2023.

Pollution also contributes to our public health climate crisis. According to researchers, human bodies are now increasingly exposed to an onslaught of dangerous contaminants in air, soil, and water. Recent testing by the Environmental Protection Agency found at about two-thirds of the U.S. population are exposed to perfluoroalkyl and polyfluoroalkyl substances (PFAS), or forever chemicals, in their tap water. As we have discussed, these man-made and persistent chemicals are used to make products heat-, water-, and stain-resistant and have been associated with such diseases as birth defects, cancer, liver disease, thyroid problems, decreased immunity, and hormone disruption in human beings (Perkins, 2024). NPR recently reported that according to recent studies, wildfire smoke now contributes to thousands of deaths each year in the United States and that accelerating climate change will likely raise this number in future years (Borunda, 2024).

One of the most important things we can do is help educate our communities about these kinds of dangers. As we all become wiser consumers and learn how to buy only products that are PFAS-free, we all help contribute to co-creating clean water. As voters, we can also urge our representatives to pass regulatory legislation with strong enforcement provisions that can help protect us from exposure to dangerous chemicals being released into our food, water, and air.

In addition to the biological vulnerabilities that our community populations have in our public health climate crisis, there are also short- and long-term psychosocial vulnerabilities. According to a study from the National Institutes of Health (Cao-Lei et al., 2022), trauma (e.g., climate-related emergencies) is associated with such symptoms as intrusive phenomena, avoidance, changes in mood and cognition, and hyperarousal in future generations. These symptoms can be transferred to those generations through a complex interaction of genetic and social events and can even be transferred when members of various generations have never actually met each other (Khan, 2024).

In general, an increasing number of people—younger adults in particular—are concerned about public mental health, and a growing number link their mental health to the climate crisis. A recent Axios-Ipsos poll reported that 17% of respondents see mental health as the biggest threat to U.S. public health, in comparison with 19% who chose obesity, 24% who identified opioids and fentanyl, and 15% who said access to guns. Mental health was the highest concern for young adults (ages 18–29) at 22%, in comparison with 10% of adults ages 65 and older (Millman, 2024).

About half of our youth report feelings of fear, sadness, powerlessness, and guilt regarding the climate crisis, and 75% see their future as "frightening." Overall, about two-thirds of the U.S. population now report that the climate crisis is important to them (Matei, 2024). And, as we reviewed in earlier chapters, in his public statements, U.S. Surgeon General Vice Admiral Vivek H. Murthy has linked the climate crisis with our mental health. As both professionals and the public recognize the impact of the climate crisis on our mental health, hopefully it will not continue to be as underassessed and undertreated as it has been during the past decades.

Although I cannot make the climate crisis go away today, I can choose to change the way I respond to it. Researchers have found that people from all over the world are experiencing such difficult emotions as helplessness, despair, fear, sadness, and anger in response to a climate crisis that seems inescapable. These feelings intensify when we have had personal experiences with climate-related events, such as a fire, flood, or heat wave. People often report, however, that they can find relief and healing by regularly going to therapy, working in a green industry, and joining with other people to respond to the climate crisis (Rudgard, 2024). It can be worrying and depressing to watch our beautiful planet become increasingly degraded, polluted, and more dangerous, but when we take positive action in response to the crisis, we can feel less helpless and more empowered.

There are many opportunities to make a difference, and when we make a difference in the world, we can do so in ourselves as well. For example, many people do not know that the United States' popular blue jeans are produced in a way that is harmful to our anthromes. Billions of denim items are made each year, with a world market of $63.5 billion recorded in 2020. The indigo dye used to color our blue jeans does not dissolve in water, so it must be combined with harsh chemicals to bind it. Those chemicals can harm textile workers (i.e., they have been found to be carcinogenic) and our environment (discharges can sometimes turn local streams blue with chemicals that can harm wildlife). Nearly 30 gallons of water are currently used to dye one pair of jeans. Fortunately, researchers in Denmark have recently found a new nontoxic enzyme that can be used in the coloring process and that can reduce environmental impacts by 73%. Helping professionals can educate our local communities about these facts and encourage citizens to only purchase blue jeans that use a safer production process (Ferrari, 2024). Then we can feel good about both wearing our new "green blue jeans" and more self-empowered as well.

At the eco and macro levels, helping professionals can contribute to the development and implementation of public- and private-sector programs designed to help prevent and treat climate-related mental health issues. We can ask for public and professional educational programs that cover topics related to public health in the climate crisis. The Environmental Protection Agency, for example, has accelerated its efforts to help disinvested communities in addressing historic climate injustices. These communities include low-income communities of color (Azhar, 2024).

At the micro and mezzo levels, by simply acknowledging the climate crisis as a factor associated with mental health, we help elevate the consciousness of people about what affects us and why. As helping professionals, we can help raise public awareness and teach new strategies for resilience. There is evidence that the well-being of our students, clients, and patients improves when they are part of involved communities, have more open conversation about the climate crisis, and stay informed about evolving local and global issues (Matei, 2024). For example, about 66% of the U.S. population report that they "rarely" or "never" discuss global warming with family and friends. However, by joining a community volunteer group that hosts public dialogues and plants trees in the local environment, individuals can experience more social cohesion, improved self-esteem, and other mental health benefits and also help make a real difference in mitigating future weather and climate disasters (Milman, 2024).

Shifting From War Economies to Peace Economies

Over the centuries, warfare and preparations for it have had a devastating impact on our anthromes. Human casualties of war include the death and wounding of both soldiers and civilians. War and its preparations contribute to the climate crisis, including global warming, pollution, and species extinction, as well as empty the treasuries of nations away from other projects that could benefit the well-being of humans, other living things, and the ecosystems that keep us all alive. Although such effects can be traced back over many centuries, impacts have accelerated as our ability to destroy has accelerated (Frankopan, 2023).

As illustrated in Table 14.5, according to the Watson Institute for International and Public Affairs at Brown University (2023), the world has seen many examples of how war harms humans since the September 11, 2001 (9/11) terror attacks. In Table 14.6, we see examples of how war and preparations for war have affected the climate crisis.

TABLE 14.5 Examples of Costs of War and Preparations for War

More than 940,000 people died in post-9/11 wars from direct war violence.
About 3.6 million to 3.8 million died indirectly in post-9/11 war zones, at a total of 4.5 4.7 million.
U.S. spending for post-9/11 wars is more than $8 trillion.
More than 432,000 civilians have died as a result of the fighting.
About 38 million people have become war refugees or displaced.
The United States conducts counterterror activities in 78 countries.
At least four times more active-duty and war veterans of post-9/11 wars died from suicide than from combat.
Wars have brought violations of human rights and civil liberties across the world.

Source: Brown University Watson Institute for International and Public Affairs, 2023.

TABLE 14.6 Examples of Ways That War and Preparations for War Impact Our Climate Crisis

Warfare directly harms wildlife and biodiversity, killing up to 90% of large animals in an area.
Pollution from war contaminates air, water, and soil to such an extent that it makes biomes unsafe for human life.
Militaries across the world create about 6% of all GHG emissions.
Figures for world militaries may be underestimated or unreported. Data are often hidden under categories such as aviation, industry, and public buildings.
Departments of defense require vast numbers of buildings, which, in the U.S. military, account for about 40% of fossil fuel consumption.

Source: Brown University Watson Institute for International and Public Affairs, 2023.

The Russian invasion of Ukraine, for example, has caused an expanding global food crises, large-scale human evacuation, and large-scale loss of human life and that of other living things. Ukraine's natural environment has experienced a tremendous loss of biodiversity and widespread damage. The war's activities have contributed to the global climate crisis through bombing of biomes and anthromes; polluting air, water, and soil; and consuming immense amounts of fossil fuels. Ukrainian activists have documented hundreds of Russian environmental crimes and called for charges of ecocide to be brought against Russia by international courts (McCarthy, 2022).

A similar impact has come from the current Israel-Hamas war as well, which is being fought in an already ecologically fragile region of the world with limited water and land resources. The U.S. military, which is estimated to emit more carbon dioxide annually than many countries, is supplying weapons to Israel. And Iran, a major supporter of Hamas, is the largest fossil fuel producer that has still not signed the Paris Agreement despite the country's domestic challenges of air pollution, droughts, and floods (Taft, 2023).

Not only do war and preparations for it make our climate crisis worse, but our national defense experts acknowledge that the climate crisis itself is a significant threat to our own national security. According to the Columbia Climate School:

> climate change is affecting practically everything on Earth, from natural systems to human endeavors. National security is no exception. The National Intelligence Council has found that "climate change will increasingly exacerbate risks to U.S. national security interests as the physical impacts increase and geopolitical tensions mount about how to respond to the challenge. The U.S. Defense Department recognizes that climate change is a "threat multiplier" as it exacerbates existing environmental stresses and security risks. In a 2021 Department of Defense report, Secretary of Defense Lloyd Austin said that almost everything the U.S. Defense Department (DOD) does to defend the American people is jeopardized by climate change—the department's strategies, plans, capabilities, missions, and equipment—and the risks are growing, especially since the world is not on track to meet its Paris Agreement goals. The risks lie not only within U.S. borders; our partner countries impacted by climate change affect American national security interests as well. (Cho, 2023)

To mitigate our climate crisis, we want to wage peace, not war. There have been attempts to create a Department of Peace within the U.S. government. On September 22, 2005, Senator Mark Dayton of Minnesota introduced Department of Peace legislation (S. 1756) that called for the expansion of former President Ronald Regan's Institute of Peace into a cabinet-level position. The U.S. Department of Peace would conduct research and establish nonviolent approaches to domestic and international conflict. The current bill is 2024 H.R. 808 (P.E.A.C.E., 2017).

We often use the metaphor of war to describe efforts we make to deal with the challenges our nation faces. We have declared a war on drugs, a war on crime, a war on poverty, and so on. These kinds of approaches tend to not work well, as they are often politically motivated and use harsh (i.e., warlike) approaches to achieve stated goals. Some may argue that there are justifications for war, such as self-defense, but such justifications can also be used to rationalize and explain a philosophy of attack.

Instead of waging a war on the climate crisis, which would justify harsh approaches to reach our goals, we can wage peace on it. A peaceful approach uses nonviolent methods to resolve conflicts. All voices are heard in the decision-making process, including those of minoritized and disadvantaged populations. This peaceful approach also strives to not only end violent conflict but also to co-create a “deep” or “positive peace.” *Positive peace* can be thought of as not merely the absence of war (which can be called *negative peace*) but also the inclusion of such positive approaches to human relationships as cooperation; the fostering of well-being, inclusion, social justice, and equity; and conflict management (e.g., Diehl, 2016).

What would a peace-based economy look like? In such an economy, international relations are based on cooperation, rather than on competition and war. In addition, there is recognition that a climate change in ecosystems can benefit all nations economically. In a peace-based economy, war funding can be increasingly diverted toward promoting the well-being of people in all nations.

Global military spending for all nations exceeded $2 trillion for the first time in 2021. Total global military expenditure increased by 0.7% in real terms in 2021, reaching $2113 billion. The five largest military budgets made up 62% of this total and were in the United States, China, India, the United Kingdom, and Russia (Stockholm International Peace Research Institute, 2022). Even a fraction of this amount of money would significantly help finance the greening of developing countries so that they could afford the expensive transition away from GHG emissions to clean energy.

Research have shown that such a transition is not only affordable but also would create more jobs and benefit the economic health of nations. An important study at Stanford University (Jacobson et al., 2019) created a blueprint for that 143 countries that account for 99.7% of the world’s planet-heating carbon pollution to switch to 100% wind, water, and solar (WWS) energy along with storage and efficiency programs by 2050. The report’s authors offer the following summary of their findings:

> Global warming, air pollution, and energy insecurity are three of the greatest problems facing humanity. To address these problems, we develop Green New Deal energy roadmaps for 143 countries. The roadmaps call for a 100% transition of all-purpose business-as-usual (BAU) energy to wind-water-solar (WWS) energy, efficiency, and storage by 2050 with at least 80% by 2030. Our studies on grid stability find that the countries, grouped into 24 regions, can match demand exactly from 2050 to 2052 with 100% WWS supply and storage. We also derive new cost metrics. Worldwide, WWS energy reduces end-use energy by 57.1%, aggregate private energy costs from $17.7 to $6.8 trillion/year (61%), and aggregate social (private plus health plus climate) costs from $76.1 to $6.8 trillion/year (91%) at a present value capital cost of ~$73 trillion. WWS energy creates 28.6 million more long-term, full-time jobs than BAU energy and needs only ~0.17% and ~0.48% of land for new footprint and spacing, respectively. Thus, WWS requires less energy, costs less, and creates more jobs than does BAU. (Jacobson et al., 2019, pp. 1–2)

In short, the hard work by the Stanford group demonstrates that fears of economic catastrophe are a much exaggerated myth and probably primarily disinformation provided by those who would benefit most from prolonging the current carbon-based economy.

However, this does not mean that financing the shift to WWS energy will be easy. In order to deploy clean energy, we will have to fund the up-front costs for industry infrastructure. According to economists, one complication for finding this funding is lies in the increase in interest rates. In addition, emerging economies especially cannot afford the cost of borrowing that would help pay up-front costs. At COP28, representatives from the Global South (Africa; Latin America and the Caribbean; Asia excluding Israel, Japan, and South Korea; and Oceania excluding Australia and New Zealand) asked for $100 billion in annual climate finance (Worland, 2023).

Antienvironmental politicians and industries often use dichotomous framing when they suggest that environmental protection means that people will lose their jobs. Unfortunately, the truth is that their information also has little merit. According to the *Scientific American*, a recent review of the peer-reviewed literature concluded that environmental regulations have had very little effect on jobs and employment in the industry being regulated (Oreskes, 2024).

A new map from the Massachusetts Institute of Technology illustrates the U.S. locations where our shift to clean energy will most affect jobs. In addition to regions that drill for oil and gas, regions with high-density manufacturing, agriculture, and construction will also be impacted, since they depend upon energy from coal, oil, and gas (Toussaint, 2024). As we move toward supporting climate justice, policymakers need to assist all workers who are affected by the decline of fossil fuels so that they are not unfairly burdened with the transition we are in.

Indeed, helping professionals can assist community members in seeing that WWS energy can bring more job opportunities to their community. For example, an energy storage company using new WWS technology is building its central manufacturing plant where a West Virginia steel mill once thrived in Weirton. This project is possible because of incentives from the Inflation Reduction Act (IRA), signed by President Biden in 2022 (Noor, 2023).

Shifting from a war economy to a peace economy means cooperating for resources rather than fighting over them and therefore is consistent with the principles of climate justice, in which we share resources equitably. For example, in our current world, where one-quarter of humanity is affected by drought and a lack of fresh water and desalination of ocean water currently requires a lot of energy (Sengupta, 2024), we can focus upon co-creating projects in which different regions, countries, and states all share in building water infrastructure and benefiting from the results. Perhaps we might be less likely to fire salvos of missiles at our neighbors if we all shared the same desalination systems that supply our thirsty populations.

It is also important to remember that if we fail to make the ecosystem climate changes necessary to reverse global warming, the economic price and cost in human suffering is likely to be immense. For example, as global warming continues, the most favorable climate for humanity is shifting toward the north and south poles, between which 600 million people currently live. As these trends continue, by the last decades of this century, an astonishing 3 billion to 6 billion people—between 33% and 50% of global population—may eventually be forced to live between those zones with increasing heat, food shortages, and mortality rates. Many of these people already live in poverty, making them especially vulnerable to these climates (Lustgarten, 2023).

Making Better Climate Solutions

Recent Goals and Progress

The International Finance Corporation Climate Center issued a report recommending several major climate solutions. The report first recommended a big increase in EVs on the road by 100 times the current levels by 2050, including not only cars and light trucks but also heavy-duty buses and trucks. The report also concluded that by 2050, almost all U.S. buildings would have to have energy efficiency and electrification updates. Regarding the U.S. power sector, the report states that we need to increase renewables to 85% of total electricity generation by 2050. This shift in power generation requires improved battery storage technologies, given how intermittent solar and wind power are (Freedman, 2023a).

We can site some important examples of climate solutions as we review our recent global progress. We know that to reach net-zero emissions by 2050, we have to decarbonize our energy production, homes and buildings, transportation, and industries. The Biden administration has, for example, made the process of creating renewable energy projects simpler, as did many U.S. states, tribes, and territories. We have increased our renewable energy capacity and are seeing more retirement of fossil fuel energy production. The Biden administration is funding both man-made carbon capture projects as well as natural projects, such as the restoration of bison, which helps our grasslands retain carbon in the soil. U.S. consumers purchased more than 1 million EVs in the United States in 2024, and the predicted number of sales by 2030 are 30 million. Hundreds of thousands of public charging ports are being built, and electric buses, passenger train lines, and bicycle routes are being constructed.

We also use a tremendous amount of energy heating, cooling, and powering homes and other buildings, and efforts are under way to make these structures more efficient so that they use much less power. Innovations such as insulated shades, exterior awnings, hempcrete (a green concrete alternative), and heat pumps are being installed. Efforts to identify and update companies that pollute are also under way, with California leading the way. Some of the most polluting industries, such as steel and concrete, still require more attention. Finally, the efforts of our children are notable. U.S. youth won the *Held v. Montana* case and are now involved in other similar legal efforts in Hawaii and California to sue governments that fail to protect the environment (Angueira, 2023).

Rewilding

What is *rewilding*? According to Rewilding.org (2024):

> rewilding is a comprehensive, often large-scale, conservation effort focused on restoring sustainable biodiversity and ecosystem health by protecting core wild/wilderness areas, providing connectivity between such areas, and protecting or reintroducing apex predators and highly interactive species (keystone species). ... The shorthand definition of Rewilding is the "3 C's"—conservation of Cores, Corridors, and Carnivores.

Since rewilding can benefit not only wilderness areas but also those heavily used by humans, we can expand this definition to include all anthromes. For example, in *Wilding*, Isabella Tree (2019) describes the rewilding work she and her husband, Charlie Burrell, did on their farm at Knepp Castle, United Kingdom. They brought in free-roaming animal herds that restored the soil just as past megafauna always have. Despite opposition from neighbors, the experiment succeeded, and the farm became a refuge for threatened species such as peregrine falcons, turtle doves, and purple emperor butterflies. Thus, their rewilding work benefited both farmers and biodiversity. Tree still advocates for farmers across the world to work on similar rewilding projects, which can help restore the soil and biodiversity while still protecting farms and farming (Ferguson, 2024).

Most scientists support the concept of rewilding but also warn that unforeseen consequences can arise when we experiment with reintroducing plants and animals into anthromes. For example, perhaps surprisingly, when people plant trees where trees have not naturally grown, harm to the environment can result. When people do so in grasslands, they can cause harm to wildlife biodiversity, local human populations, the water cycle, and even local climate (Leffer, 2024).

Helping professionals can help create and facilitate dialogues in which people learn from experts about rewilding and discuss what they feel and think about proposed projects in their local community.

Wind/Water/Solar and Other Alternative Energy Sources

As technology improves and costs decrease, WWS power will be increasingly available in our communities.

Researchers at Chalmers University of Technology in Gothenburg, Sweden, have invented technology that can capture and store solar energy for up to 18 years and produce electricity when needed. Such technology will enable solar energy to be stored, sent to other locations, and converted into electricity when needed (Budin, 2024).

Hydrogen is found abundantly on Earth and is a primary element found in stars. It can carry and store a great deal of energy but is not an energy source. Hydrogen is currently used in fuel cells to generate electricity, power, or heat. Today, hydrogen is most often used in petroleum refinery and fertilizer production, whereas transportation and utilities are emerging markets. Hydrogen is also used on spacecraft for propulsion and electrical power. Ultimately, it can potentially help decarbonize many industries and reduce more than 20% of annual global emissions by 2050. It is expensive to produce hydrogen because it takes energy to do so. The big challenge for hydrogen production, particularly from renewable resources, is to produce it at a lower cost (Gibson, 2023, p. 1).

Identifying New Partners

One of my favorite climate solutions is to train and hire local displaced workers to be the caretakers of local anthromes. In December 2023, Colombia created a new national park near its border with Venezuela near a tributary of the Amazon River. The government has employed displaced rangers to be the rangers of this new park (Smith, 2024). It turns out that these ranchers are also excellent rangers. Who would have been better to help protect the land than people who have worked on it for decades and love it as much as any of the urban environmentalists in the country?

Similarly, we also may be able to find wildlife partners who can help us reduce GHGs, reduce pollution, and promote biodiversity.

For example, scientists have found viruses that may help us decrease the level of GHGs in our air and water. About 128 viruses have been discovered so far. Some eat carbon in seawater, and others stop methane from escaping out of melting permafrost. Ohio State University microbiologist Matthew Sullivan and colleagues use artificial intelligence (AI) modeling and genome sequencing to find these viruses (Al-Sibai, 2024).

Geoengineering and Climate Technology

Some experts insist that it is not enough for humanity to try to cut back on our GHG emissions and that we also need to directly remove carbon from our water and air. Some researchers are exploring ways to reduce GHGs by creating large-scale engineering projects.

A recent article in *The Wall Street Journal* describes three geoengineering projects that aim to change atmospheric or ocean chemistry to lessen the effects of GHG emissions. These include adding chemicals to the ocean, saltwater to clouds, and reflective particles to the sky. Unintended consequences of these trials are unknown (Niiler, 2024).

Climate tech or *clean tech* industries make up a newly developing sector that uses widespread applications from creating vegetarian meat to smart mining. Some experts (Taft, 2024) are concerned that these emerging industries are often heavily funded by fossil fuel money, from corporations that originally contributed to the GHG problem. Such funding could be used to steer these new climate tech industries from working on projects that could directly compete with the profitable operations of those fossil fuel corporations. Experts are concerned that if we allow the market to serve as the solution for our climate crisis, we seem likely to continue to hear little critique of the "capitalist framework that drives tech attitudes" (Taft, 2024).

We may need to protect climate tech from our capitalistic economic framework. Although private-sector funding can be a valuable addition to funding from government, there are dangers in allowing corporations to have too much influence, particularly when their own profits are being threatened. The International Monetary Fund warned in 2023 that public funds will not by themselves be sufficient to pay for the installation of clean energy (Freedman, 2023b).

Regularly Reviewing Outcomes and Updating Solutions

As we try new policies to protect our anthromes, we can also learn more about what works and what needs to be updated or changed.

For example, bans can work but usually have flaws that need to be corrected. Plastic pollution, as we have discussed, has become dangerously ubiquitous across the world. Single-use plastic bag bans have helped, but compromises in new laws that were seen as necessary for public support have created new problems. Single-use plastic bag bans can reduce use by 300 bags a year, and nationwide we have reduced consumption by about 6 billion bags per year. However, people need to purchase multiple use bags and use them longer in order to reduce unnecessary waste and pollution. In order to correct some of the imperfections in plastic bag bans, consumers are recommended to use paper bags four to eight times and high-density polyethylene bags five to 10 times before disposing of them. Durable tote bags should be used one to 20 times, and cotton bags should be used 50 to 150 times (Weise, 2024).

Another example of a promising climate solution involves our seafood. By ending overfishing, we can protect fish biodiversity, strengthen marine life resilience, and reduce carbon emissions. Large subsidized fishing fleets are associated with depleted resources and increased emissions. Fish populations can rebound with regulations in place. Ecosystem well-being should be prioritized over providing the maximum number of fish products for profit. Fish also suffer from ocean warming and heavy metal pollution, which reduces their reproductive success, increases disease, and allows invasive species to enter vulnerable biomes. International treaties that regulate overfishing need to be designed and ratified. When we protect the ocean, it can better protect us by helping feed us and absorb carbon safely (Sumaila, 2024).

Considering the Risks and Benefits of Using AI in Addressing the Climate Crisis

University of Cambridge Professor of Politics David Runciman (2023) offers an alternative view of the macro-level solutions available to us in addressing such current global challenges as nuclear war, climate crisis, and biological catastrophe. He argues that the outcomes of major climate solutions are currently out of our hands and instead depend on how the states and corporations we have created will use and regulate the emerging AIs we are currently creating. He calls states and corporations "artificial agencies" (AAs) that have been developed by human intelligence (HI).

Climate activists are already using AI technologies. For example, a recent article in *The Guardian* described how a group of young volunteers who are translating climate information from English into other languages used across the world are now using new Google-designed AI tools to accomplish their goals. These AI tools help the volunteers reach their objectives more quickly and accurately (Turns, 2023).

Runciman offers a list of factors that can help inform what choices we make. As you read the list, you can start to see the complexities of the choices that we are now investigating. If the "smartest" humans are in charge of the state, democracy suffers, and elitism can result. AI can outperform humans in many tasks but currently relies on humans to choose them. So, if intelligent politics need to be task-centered, this favors corporations, which currently lead the world in the use of task-centered AI applications. If nonhuman agents (i.e., AAs) are put in charge, humans will be tempted to avoid accountability and to blame AAs instead. If states, corporations, or robots can protect us, we could strive to make them smarter. The state is stronger and more durable than corporations, but corporations can sometimes avoid being regulated by states. States remain "vehicles of human choice," but human choices can sometimes be the wrong ones. Finally, since states often struggle to regulate corporations, how can we expect states to regulate smarter AIs effectively?

In column 1, we see the three choices that Runciman states are available to humanity. The choice "HI + AI," for example, involves giving agency to a combination of HI and AI. In column 2, we see a simple comparison of the answers we would obtain from each choice. Similarly, accountability outcomes are described in column 3, and what we are calling *inclusivity outcomes* are compared in the last column.

TABLE 14.7 Humanity's Three Choices

Three Choices	Answer Outcomes	Accountability Outcomes	Inclusivity Outcomes
HI + AI	Better answers	Worse accountability	Elitism
AA + AI	Better answers	Worse accountability	Inhumanity
AA + HI	Worse answers	Better accountability	More human

Source: Runciman, 2023.

What are the implications of Runciman's work for our macro-level efforts to address the climate crisis? Whether we agree with him that we have handed over control of our lives to corporations, states, and AIs or not, we can still heed his warning and strive to find the wisest system of checks and balances between HI, AA, and AI. Perhaps the ideal system would both incorporate the strengths of HI, AA, and AI while mitigating the limitations and dangers of each.

QUESTIONS FOR REFLECTION

1. This chapter offered a large and diverse collection of climate issues and solutions. What was your experience about these issues like? How did your feeling and thoughts about your world change, if at all?
2. Please review Table 14.2, adapted in the "Practice Topics" table. What topics are you the most interested in? Do you know why? Please explain.
3. After having read about the first topic throughout the text, how do you assess your own faith-based activism?
4. With which elements of activism wisdom did you resonate the most? Please explain why.
5. To what extent have you been affected by the polarization and outrage spread by media technology? Please explain.
6. In what locations in the world would you choose to live, if avoiding weather and climate disasters was a priority for you?
7. How has the climate crisis affected the public health of the community in which you live now?
8. What was your reaction to the ideas expressed in the section about peace economies?
9. Did you have a favorite climate solution? Please explain.

Practice Topics
Fostering faith-based activism
Developing activism wisdom
Mitigating polarization and outrage
Mitigating the impacts of weather and climate disasters
Mitigating our public health climate crisis
Shifting from war economies to peace economies
Making better climate solutions

REFERENCES

Al-Sibai, N. (2024). *Scientists propose hacking viruses to fight climate change*. Futurism, February 19. 2024. https://futurism.com/the-byte/climate-change-pollution-viruses

Angueira, G. A. (2023). *A look back at U.S. climate solutions this year*. Grist, December 21, 2023. https://grist.org/solutions/a-look-back-at-u-s-climate-solutions-this-year/

Anthroecology Lab. (2023). *Guide to anthromes: Explore by anthrome*. Anthroecology. https://anthroecology.org/anthromes/guide/anthrome/

Arigoni, D. (2023). *Climate resilience for an aging nation.* Island Press.

Azhar, A. (2024). *A year before Biden's first term ends, environmental regulators rush to aid disinvested communities.* Inside Climate News, February 6, 2024. https://insideclimatenews.org/news/06022024/epa-rush-to-aid-disinvested-communities-before-end-of-biden-first-term/

Biden, J. (2022). *Remarks by President Biden on actions to tackle the climate crisis.* WhiteHouse.gov, July 20, 2022. https://www.whitehouse.gov/briefing-room/speeches-remarks/2022/07/20/remarks-by-president-biden-on-actions-to-tackle-the-climate-crisis/

Borunda, A. (2024). *Wildfire smoke contributes to thousands of deaths each year in the U.S.* National Public Radio, April 18, 2024. https://www.npr.org/2024/04/18/1245068810/wildfire-smoke-contributes-to-thousands-of-deaths-each-year-in-the-u-s

Brown University Watson Institute for International and Public Affairs. (2023). *Costs of war: Summary of findings.* https://watson.brown.edu/costsofwar/papers/summary

Budin, J. (2024). *Researchers achieve breakthrough in solar technology: "A radically new way of generating electricity from solar energy."* The Cool Down, February 25, 2024. https://www.thecooldown.com/green-tech/solar-energy-storage-system-sweden/

Cao-Lei, L., Saumier, D., Fortin, J., & Brunet, A. (2022). A narrative review of the epigenetics of post-traumatic stress disorder and post-traumatic stress disorder treatment. *Frontiers in Psychiatry, 13,* 857087. https://www.ncbi.nlm.nih.gov/pmc/articles/PMC9676221/

Carnell, H. (2024). A bunch of new research puts this winter's wild weather in frightening context. *Mother Jones,* February 9, 2024. https://www.motherjones.com/environment/2024/02/climate-change-science-record-winter-storm/

Cho, R. (2023). *Why climate change is a national security risk.* State of the Planet, October 11, 2023. Columbia Climate School. https://news.climate.columbia.edu/2023/10/11/why-climate-change-is-a-national-security-risk/

Cohen, L. (2023). *World's richest 1% emitting as much carbon to cause heat-related deaths for 1.3 million people, report finds.* CBS News, November 21, 2023. https://www.cbsnews.com/news/worlds-richest-carbon-emissions-climate-change-report/?ftag=CNM-00-10aac3a

Coren, M. (2023). The surprisingly simple way to convince people to go green. *The Washington Post,* December 5, 2023. https://www.washingtonpost.com/climate-environment/2023/12/05/improve-sustainability-help-climate-change/

Diehl, P. F. (2016). Exploring peace: Looking beyond war and negative peace. *International Studies Quarterly, 60*(1), 1–10. https://doi.org/10.1093/isq/sqw005

Ellis, E. C. (2020). *Anthromes.* Encyclopedia of the world's biomes. Anthroecology. https://anthroecology.org/wp-content/uploads/2020/09/ellis_2020a.pdf

Feiger, E. (2023). *New analysis reveals which state should be most concerned about worsening extreme weather: "It's a domino effect."* The Cool Down, December 27, 2023. https://news.yahoo.com/one-world-biggest-cities-may-103023024.html

Ferguson, D. (2024). "Does rewilding sort climate change? Yes!": U.K. expert says nature can save planet and not harm farming. *The Guardian,* February 25, 2024. https://www.theguardian.com/environment/2024/feb/25/rewilding-climate-change-biodiversity-isabella-tree-nature-planet-farming?CMP=oth_b-aplnews_d-1

Ferrari, O. (2024). Blue jeans are terrible for the environment—but a new discovery could help. *National Geographic,* February 27, 2024. https://www.nationalgeographic.com/environment/article/blue-jeans-indigo-indican-enzyme

Frankopan, P. (2023). *The Earth transformed: An untold story.* Knopf.

Freedman, A. (2023a). *U.S. could reach net-zero target through these key steps: Report.* Axios, October 4, 2023. https://www.axios.com/2023/10/04/us-net-zero-emissions-2050

Freedman, A. (2023b). *Paris targets need big private climate spending boost: IMF.* Axios, October 3, 2023. https://www.axios.com/2023/10/03/imf-clean-energy-climate-spending

Freedman, A. (2024). *Climate change may cost $38 trillion a year by 2049, study says.* Axios, April 18, 2024. https://www.axios.com/2024/04/18/climate-change-damages-38-trillion

Friedman, L. (2023). Countries most at risk call proposed climate agreement a "death warrant." *The New York Times,* December 12, 2023. https://www.nytimes.com/2023/12/11/climate/cop28-climate-agreement-un.html#:~:text=agreement%2Dun.html-,Countries%20Most%20at%20Risk%20Call%20Proposed%20Climate%20Agreement%20a%20'Death,are%20dangerously%20heating%20the%20planet

Gibson, K. (2023). *What is hydrogen energy, and is it a key to fighting climate change?* CBS News, October 18, 2023. https://www.cbsnews.com/news/hydrogen-energy-what-is-climate-change/?ftag=CNM-00-10aac3a

Glueck, K. (2023). Anti-Trump burnout: The resistance says it's exhausted. *The New York Times,* February 19, 2023. https://www.nytimes.com/2024/02/19/us/politics/trump-resistance-democrats-voters.html

Harvey, F. (2024). Demagogues imperilling [sic] global fight against climate breakdown, says Kerry. *The Guardian*, February 28, 2024. https://www.theguardian.com/environment/2024/feb/28/populism-imperilling-global-fight-against-climate-breakdown-says-john-kerry?CMP=oth_b-aplnews_d-1

Haupt, A. (2024). 7 ways to deal with climate despair: Here's how climate-aware therapists help their clients cope. *TIME*, February 6, 2024. https://time.com/6589649/climate-despair-how-to-cope/

Hughes, R. A. (2023). *Climate Now debate: 2023 is set to be the hottest year on record, so why aren't we taking action?* euronews, October 13, 2023. https://www.euronews.com/green/2023/10/13/climate-now-debate-2023-is-set-to-be-the-hottest-year-on-record-so-why-arent-we-taking-act

Jacobson, M. Z., Delucchi, M. A., Cameron, M. A., Manogaran, I. P., Shu, Y., & von Krauland, A. (2019). Impacts of Green New Deal energy plans on grid stability, costs, jobs, health, and climate in 143 countries. *One Earth Journal*, *1*(4), 449–463. https://www.cell.com/one-earth/fulltext/S2590-3322(19)30225-8

Kaufman, L. (2023). *More than 3 million Americans are already climate migrants, researchers say.* Bloomberg, December 18, 2023. https://www.bloomberg.com/news/articles/2023-12-18/more-than-3-million-americans-are-already-climate-migrants-researchers-say

Kendzior, S. (2022). *They knew: How a culture of conspiracy keeps America complacent.* Flatiron.

Khan, U. (2024). *Oh good—you can inherit your grandparents' trauma, even if you've never met them.* HuffPost, February 19, 2024. https://www.huffingtonpost.co.uk/entry/your -grandparents-trauma-can-make-you-anxious-even-if-youve-never-met -them_uk_65d32189e4b043f1c0abdd96?ncid=APPLENEWS00001

King, M. L., Jr. (2012). *A gift of love: Sermons from strength to love and other preachings.* Beacon Press.

Leffer, L. (2024). When planting trees is bad for the planet. *Popular Science*, February 15, 2024. https://www.popsci.com/environment/when-planting-trees-is-bad-for-the-planet/

Lustgarten, A. (2023). *Climate crisis is on track to push one-third of humanity out of its most livable environment.* ProPublica, June 6, 2023. https://www.propublica.org/article/climate-crisis-niche-migration-environment-population

Marino, J. (2024). *Researchers uncover potential key to combating non-communicable diseases—and it's surprisingly simple.* The Cool Down, February 21, 2024. https://www.thecooldown.com/green-tech/plant-based-diets-benefits-sustainable-futures/

Matei, A. (2024). Climate anxious? Here's how to turn climate anxiety into action. *The Guardian*, February 27, 2024. https://www.theguardian.com/wellness/2023/nov/16/climate-anxiety-tips?CMP=oth_b -aplnews_d-1McCarthy, J. (2022). *How war impacts climate change and the environment.* Global Citizen, April 6, 2022. https://www.globalcitizen.org/en/content/how-war-impacts-the-environment-and-climate-change/

McKay, D.I.A., et al. (2022). Exceeding 1.5°C global warming could trigger multiple climate tipping points. *Science*, Aug., https://www.science.org/doi/10.1126/science.abn7950

Milman, O. (2024). Study: Reforestation has protected eastern states from temperature rise. *Mother Jones*, February 23, 2024. https://www.motherjones.com/environment/2024/02/study-trees-cooling-protecting-eastern-states/

Millman, J. (2024). *Mental health seen as a top health threat in Axios-Ipsos poll.* Axios, February 23, 2024. https://www.axios.com/2024/02/23/mental-health-threat-poll-drugs-guns

Mulvaney, K. (2023). What's the big deal about Earth getting 2°C hotter? *National Geographic*, December 1, 2023. https://www.nationalgeographic.com/environment/article/paris-climate-agreement-earth-two-degrees-hotter

National Centers for Environmental Information. (2024). *Billion-dollar weather and climate disasters: Overview.* March 8, 2024. https://www.ncei.noaa.gov/access/billions/

Newcomb, T. (2024). Turns out Earth's temperature timeline is wrong. *Popular Mechanics*, February 15, 2024. https://www.popularmechanics.com/science/a46769291/turns-out-earths-temperature-timeline-is-wrong/

Niiler, E. (2024). Scientists resort to once-unthinkable solutions to cool the planet. *The Wall Street Journal*, February 15, 2024. https://www.wsj.com/science/environment/geoengineering-projects-cool-planet-weather-f0619bf7

Noor, D. (2023). Renewable energy jobs are coming to Appalachia. *Mother Jones*, August 22, 2023. https://www.motherjones.com/politics/2023/08/renewable-energy-jobs-are-coming-to-appalachia/

Oreskes, N. (2024). Environmental protection does not kill jobs. *Scientific American*, February 1, 2024. https://www.scientificamerican.com/article/environmental-protection-does-not-kill-jobs/

P.E.A.C.E. (2017). *Create a department of peacebuilding.* October 29, 2017. https://peaceeducators.org/resources/create-department-peacebuilding/

Perkins, T. (2024). At least 60% of U.S. population may face "forever chemicals" in tap water, tests suggest. *The Guardian*, February 20, 2024. https://www.theguardian.com/environment/2024/feb/20/pfas-us-drinking-water-tap?CMP=oth_b-aplnews_d-1

Price, K. (2024). *Atmospheric rivers in California create a perfect storm of public health risks.* Climate News, February 20, 2024. https://insideclimatenews.org/news/20022024/todays-climate-atmospheric-rivers-in-california-create-perfect-storm-of-public-health-risks/

Pulver, D. V. (2023). The weather is getting cold. Global warming is still making weather weird. *USA Today*, December 13, 2023. https://www.usatoday.com/story/news/nation/2023/12/12/climate-change-effects-in-winter/71670349007/

Reuters. (2024). *What a second Trump presidency could mean for U.S. energy policy.* Reuters, February 16, 2024. https://www.reuters.com/world/us/what-second-trump-presidency-could-mean-us-energy-policy-2024-02-16/

Ritchie, H. (2024). *Not the end of the world: How we can be the first generation to build a sustainable planet.* Little, Brown Spark.

Ritchie, H. (2023). The big idea: Why climate tribalism only helps the deniers. *The Guardian*, July 10, 2023. https://www.theguardian.com/books/2023/jul/10/the-big-idea-why-climate-tribalism-only-helps-the-deniers?CMP=oth_b-aplnews_d-1

Rodriguez, A. (2024). Americans are lonely and it's killing them. How the U.S. can combat this new epidemic. *USA Today*, March 23, 2024. https://www.usatoday.com/story/news/health/2023/12/24/loneliness-epidemic-u-s-surgeon-general-solution/71971896007/

Rose-Stockwell, T. (2023). *Outrage machine: How tech amplifies discontent, disrupts democracies—and what we can do about it.* Legacy Lit.

Rudgard, O. (2024). *Climate anxiety can feel like "there's no safe harbor."* Bloomberg, February 16, 2024. https://www.bloomberg.com/news/articles/2024-02-16/climate-anxiety-can-feel-like-there-s-no-safe-harbor-reader-survey-finds

Runciman, D. (2023). *The handover: How we gave control of our lives to corporations, states and AIs.* Liveright.

Sattler, L. (2024). *Concerning new report warns that a public health crisis is claiming more and more lives: 'It's disheartening.'* The Cool Down, February 20, 2024. https://www.thecooldown.com/green-tech/pakistan-climate-change-health-impacts/

Sengupta, S. (2024). Drought touches a quarter of humanity, U.N. says, disrupting lives globally. *The New York Times*, January 11, 2024. https://www.nytimes.com/2024/01/11/climate/global-drought-food-hunger.html#:~:text=The%20United%20Nations%20estimates%20that,%2D%20and%20middle%2Dincome%20countries

Simon, J. (2024). *Climate scientist Michael Mann wins defamation case against conservative writers.* National Public Radio, February 8, 2024. https://www.npr.org/2024/02/08/1230236546/famous-climate-scientist-michael-mann-wins-his-defamation-case

Smith, J. E. (2024). On the plains, ranchers are now Rangers. *The New York Times*, January 16, 2024. https://www.nytimes.com/2024/01/16/science/colombia-park-manacacias.html

Sommer, L. (2023). *Three climate impacts the U.S. will see if warming goes beyond 1.5 degrees.* National Public Radio, November 29, 2023. https://www.npr.org/2023/11/29/1214858764/3-climate-impacts-the-u-s-will-see-if-warming-goes-beyond-1-5-degrees

Stockholm International Peace Research Institute. (2022). *World military expenditure passes $2 trillion for first time.* April 25, 2024. https://www.sipri.org/media/press-release/2022/world-military-expenditure-passes-2-trillion-first-time

Sumaila, R. (2024). *Eight ways that stopping overfishing will promote biodiversity and help address climate change.* The Conversation, February 15, 2024. https://theconversation.com/8-ways-that-stopping-overfishing-will-promote-biodiversity-and-help-address-climate-change-218977

Taft, M. (2023). Endless war on a dying planet. *New Republic*, October 2, 2023. https://newrepublic.com/post/176184/israel-palestine-endless-war-dying-planet

Taft, M. (2024). I saw the future of climate technology—and its big-oil investors. *The Nation*, February 15, 2024. https://www.thenation.com/article/environment/cleantech-conference-climate/

Thompson, A. (2023). The most important climate stories of 2023 aren't all bad news. *Scientific American*, December 27, 2023. https://www.scientificamerican.com/article/the-most-important-climate-stories-of-2023-arent-all-bad-news/

Toussaint, K. (2024). This map shows where the shift to clean energy will most affect jobs. *Fast Company*, February 5, 2024. https://www.fastcompany.com/91022799/this-map-shows-where-the-shift-to-clean-energy-will-most-affect-jobs

Tree, I. (2019). *Wilding.* New York Review Books. https://www.nyrb.com/products/wilding

Turns, J. (2023). "The change in pace is crazy": AI boosts climate information translation drive. *The Guardian*, June 6, 2023. https://www.theguardian.com/environment/2023/jun/06/climate-cardinals-ai-boost-artificial-intelligence?CMP=oth_b-aplnews_d-1

van der Zee, B. (2024). *Not the end of the world* by Hannah Ritchie review—An optimist's guide to the climate crisis. *The Guardian*, January 4, 2024. https://www.theguardian.com/books/2024/jan/04/not-the-end-of-the-world-by-hannah-ritchie-review-an-optimists-guide-to-the-climate-crisis

Vega, P. (2023). The state of the climate crisis. *Vox*, November 27, 2023. https://www.vox.com/2023/11/27/23959446/cop28-united-nations-climate-crisis

Weise, E. (2024). Plastic bag bans have spread across the country. Sometimes they backfire. *USA Today*, February 17, 2024. https://www.usatoday.com/story/news/nation/2024/02/17/plastic-bag-bans-can-increase-or-reduce-plastic-use-heres-why/72522792007/

Worland, J. (2023). Why finance will be key to the climate story in 2024. *TIME*, December 22, 2023. https://time.com/6549989/climate-finance-cop29/

Yoder, K. (2023a). *Four in 5 people around the world support "whatever it takes" to limit climate change*. Grist, November 30, 2023. https://grist.org/language/4-in-5-people-around-the-world-support-whatever-it-takes-to-limit-climate-change/

Yoder, K. (2023b). Why fake news about climate change is still so effective. *Mother Jones*, December 24, 2023. https://www.motherjones.com/environment/2023/12/why-fake-news-about-climate-change-is-still-so-effective/

Youmshajekian, L., & Fragapane, F. (2024). Visualizing climate disasters' surprising cascading effects. *Scientific American*, February 1, 2024. https://www.scientificamerican.com/article/visualizing-climate-disasters-surprising-cascading-effects/

Young, C. (2023). Climate change isn't a top issue for Democrats or Republicans. Record heat should change that. *USA Today*, July 15, 2023. https://www.usatoday.com/story/opinion/2023/07/15/heat-warnings-climate-change-hot-topic-politics/70407209007/

Credits

CHAPTER 15

IMG 15.1

Conscious Climate Change

A quasi-religious transformation leading to the appreciation of diversity for its own sake, apart from the obvious direct benefits to humanity, may be required to save other organisms and ourselves.

—PAUL EHRLICH, *BIODIVERSITY* (1988)

The doctrine of the Incarnation is itself an invitation to all believers to love the earth, cherish it, find the divine in it.

—MATTHEW FOX, *ORIGINAL BLESSING* (GOODREADS, 2024)

Two of us riding nowhere . . .
We're on our way home.

—"TWO OF US" (LENNON & MCCARTNEY, 1970)

Once upon a time, there was Consciousness.

And Consciousness imagined a universe of many conscious beings, who were all parts of Consciousness.

And Consciousness imagined a planet. The planet Earth would be a beautiful conscious jewel in a conscious stunning universe with no apparent end. Consciousness imagined the planet in orbit around a stable yellow star, at just the right distance from Earth that its heat could support life.

The home star was one of hundreds of billions of stars in a galaxy of conscious stars. This home galaxy was one of hundreds of billions of conscious galaxies peering like sparkling eyes from the surfaces of colossal bubbles of galaxies that each encircled vast conscious voids.

The home planet itself was populated with conscious living things. The atmosphere would provide just the right balance of elements that would enable both plants and animals to thrive. The daytime color would be sky blue with a mixture of ever-changing cloud formations; the night sky would be dark enough to allow the light of the moon, planets, stars, and galaxies to reach the surface.

All the beings of the Universe were aware of the Consciousness that had imagined them into existence. And each being, whether a rock, comet, tree, or hawk, all experienced connection with Consciousness, each through their own individual consciousness. All of Creation, therefore, felt at Home in the Universe.

Then Consciousness imagined a new species with a unique purpose. These human beings were also part of Consciousness, like everything else in the Universe. However, they did not yet have the natural awareness of Creation like other beings had and thus felt disconnected from the Universe. Their purpose was to remember their connection with Consciousness.

However, Consciousness created conditions that made that life purpose meaningful. First, Consciousness imagined the planet Earth as an especially challenging place for these humans to live on, with both intense beauty and intense suffering to learn from. Second, humans were given a challenging ego that sought connection through identification with such short-lived frivolous possessions as power, wealth, knowledge, and fame.

Given a free will, humans were allowed to wage war on nature and on each other, and in doing so, they gradually destroyed much of the planet's wildlife and ecosystems, including many of their fellow human beings. However, the more the ego was successful in winning its wars and accumulating power, wealth, and knowledge, human beings felt even more dissatisfied. And even when ego was successful in finding fame and therefore believing that it had conquered time, all human bodies continued to eventually die anyway.

Human being also were given an individual and collective capacity for reflection, loving kindness, and transformation. Scattered throughout human history are stories of courageous people who led the way in rediscovering human reconnection with Consciousness and the Universe. They needed to be courageous because their messages of connection with Consciousness threatened the people in power, who were still servants of the ego's desire for wealth, power, knowledge, knowledge, and fame.

Human beings are thus all born feeling alone, and most die feeling alone. In a seemingly infinite Universe of time and space, the first modern humans beings evolved in Africa about 300,000 years ago.

We have been trying to come home ever since.

I began this last chapter in our text with this creation myth, which includes archetypes borrowed from the creation myths found in many cultures. A creation myth is a symbolic story about the origin mysteries of the Universe and human life. The purpose of the creation myth is not necessarily to explain these mysteries, but rather to help us cooperate together in supporting the well-being of all people, other living things, and the ecosystems that support all life.

In the creation myth, we use the big-C word, *Consciousness,* to describe a universal and perhaps infinite awareness that underlies the experience and existence of all beings. From this perspective, Consciousness is available not only to human beings, but to all living things, as well as even to rocks, rivers, clouds, and stars. As we have throughout our text, we can continue to also use the little-c word *consciousness* to describe the limited but reverent awareness that we humans can experience.

In Section 1, we explored how human consciousness is related to the climate crisis and climate solutions. In Chapter 15, at the end of this text, we return to consciousness again and examine how a climate change in our human consciousness may likely be the foundation of a sustainable climate change in our human systems and ecosystems.

Although historians tell us that our religiosity and spirituality has always been changing, we live in a time when our religious and/or spiritual beliefs and practices seem especially in flux. Recently we see increasing numbers of people leaving organized religion and a growing percentage also report that they are now are spiritual. Today, more than half of the U.S. population now report that they are *both* religious and spiritual. Researchers have also found that although we still have many disagreements about such topics as the identity of God, the meaning of sacred texts, and the importance of attendance in religious services, there is also growing *consensus* in the United States about how we view other living things and ecosystems (Alper et al., 2023; Lipka & Gecewicz, 2017).

Remember that we introduced *spirituality* as an individual dimension of human development, like emotional growth or physical growth, in which the person becomes more aware of their connection with the world. Spiritual development may be linked with the development of a personal life purpose and meaning. In contrast, we defined *religiosity* as a social function, in which people share beliefs, rituals, and doctrines with other people in their community.

I have chosen to use the word *consciousness* (rather than primarily using the words *spirituality* and *religiosity*) throughout this text for a number of reasons. First, the term *spirituality* is *inclusive* and can be used to describe the reverent awareness that all humans can have. The term also seems to be used today by increasing numbers of scientists, religious leaders, spiritual teachers, and the general public. In contrast, although the connotations of spirituality and religiosity may be moderating, many of us still harbor negative stereotypical associations with these terms. For example, so-called spiritual people might be viewed by some as being spacy, immoral, and self-centered, whereas so-called religious people might be viewed by some as rigid, intolerant, and controlling. Finally, in the United States today, religiosity and spirituality both appear to be associated with voting patterns, and our current political polarization can still sometimes overwhelm the emerging agreements across our faith and spiritual communities.

A Greening Convergence of Science and Spirituality

I chose the Paul Ehrlich and Matthew Fox quotes to lead the chapter because, although one is a scientist and the other a priest, they agree that that climate crisis requires spiritual and religious transformation. Biologist Paul Ehrlich (1968) authored his then controversial *The Population Bomb* in 1968, in which he predicted that accelerating global human population growth was leading us toward the accelerating climate crisis that we are dealing with today. Fox is a priest, activist, and author who has created a new vision of green spirituality and green religiosity. He has authored many books, including his influential *Creation Spirituality* (1991), which links the spiritual poverty, environmental destruction, and other social justice issues of our era.

According to Fox, in order to address our climate crisis, Western culture now needs to reinvent new creation myths and rituals, renew our faith in mystical experience, and co-create a spirituality of compassion that informs social justice activism. Fox, who calls himself a "spiritual theologian," can appeal to many people who identify as being either spiritual or religious or both. His life work calls for a greening of both spirituality and religiosity (Fox, 2024).

What would make both a renowned scientist and a popular priest, who are each concerned about our climate crisis, call for a new global religious and spiritual movement? We have seen that that many if not most humans today feel disconnected from each other and from other living things and ecosystems as well. We showed in Section 1 how the human ego contributes to that sense of separateness, as we identify with our own beliefs, culture, power, and wealth. Such identifications tend to lead to the formation of in-groups and out-groups, which can complicate our willingness to cooperate for the highest good of all. Both Ehrlich and Fox believe that an emerging green religious and spiritual consensus could help humanity rediscover our interconnectedness and thereby recommit to a new era of global cooperation in developing climate solutions.

The Green Convergence of Religion and Spirituality

It may well turn out that humanity's successful and timely response to our climate crisis will significantly depend upon growing support from people engaged in spiritual work and religious traditions:

> It is increasingly acknowledged that our escalating global environmental challenges cannot be adequately responded to by governmental bodies without a substantial upsurge of popular support for the far-reaching policy shifts widely seen as necessary. The search is thus on for the societal, cultural and moral energies that might inspire the large-scale consciousness-raising and mobilization required for such an upsurge. More recently, there has also been a growing recognition that, in a world where four-fifths of the population adheres to some form of spiritual tradition, religion may be one of those necessary sources. (Chaplin, 2016, p. 1)

Remember that there are four identity groups we can create regarding spirituality and religiosity. These are illustrated in Table 15.1, along with changing percentages reported over time as well as additional data on beliefs regarding spirits and spiritual energy in nature. The data suggest that although the relative order of the size of the four groups has remained consistent since 2012, the percentages have shifted somewhat. In column 6, we also can view percentages of respondents who reported that they believe that nonhuman animals (e.g., birds or fish) or parts of nature (e.g., rocks or rivers) have either spirits or spiritual energy. It is interesting to note that the group *least* likely to see spirits or spiritual energy in nature were the religious-but-not-spiritual group, whereas the most likely to was the spiritual-but-not-religious group. One wonders if and how these percentages will change over another decade of polling.

TABLE 15.1 Religion and Spirituality

Groups	Four Categories	2012	2017	2023	What percentage believe nonhuman animals and what percentage believe parts of nature have spirits or spiritual energy?
Group 1	Both religious and spiritual	59%	48%	48%	54% nonhuman animals*; 45% parts of nature*
Group 2	Spiritual but not religious	19%	27%	22%	78% nonhuman animals; 71% parts of nature
Group 3	Neither spiritual nor religious	16%	18%	21%	44% nonhuman animals; 35% parts of nature
Group 4	Religious but not spiritual	6%	6%	10%	54% nonhuman animals*; 45% parts of nature*

*These 2023 percentages are for groups 1 and 4 combined.

Sources: Data from two different Pew Research Center articles: Alper et al., 2023; Lipka & Gecewicz, 2017.

I believe that the number of people who are concerned about the climate crisis and willing to cooperate with others to support climate solutions will continue to grow both nationally and globally over the next decade in all four groups illustrated in Table 15.1. These shifts in public opinion are likely to occur as the intensity and frequency of climate disasters continue to increase. In addition, religious institutions are likely to increasingly ally with open-minded political leaders and the scientific community by providing the spiritual and ethical leadership required in their local and global communities. As Dasgupta and Ramanathan (2014) stated, "Finding ways to develop a sustainable relationship with nature requires not only engagement of scientists and political leaders, but also moral leadership that religious institutions are in a position to offer."

Although they were often a minority, there have been religious leaders throughout history who have expressed concern about the well-being of nonhuman life and ecosystems. We know that many ancient people seemed to have had a spiritual connection with other living things and with Earth. For example, in his book *Hildegard of Bingen: A Saint for Our Times,* Fox (2012) remembers Saint Hildegard, whom Pope Benedict XVI formally canonized in 2012. Despite the marginalization of women in the 12th century, Hildegard was a visionary who often spoke of humankind's spiritual relationship with Earth.

Interest in ecology started to grow in religious communities in the 1970s in the Americas, South Asia, and Africa (Gottlieb, 2006). Leaders from the world's major religions expressed concern together about the climate crisis at the 2015 United Nations Climate Change Conference in Paris (Chaplin, 2016). Information on the ecological concerns of our world wisdom traditions has been compiled at the Yale Forum on Religion and Ecology (https://fore.yale.edu/). The forum was founded at the United Nations in 1998 and has been based at Yale University since 2006. In 2023, the forum began working with the Yale Center for Environmental Justice.

Jonathan Chaplin has identified a list of seven "prominent concerns" or "convergences" that the major religions of the world now seem to share (see Table 15.2). Helping professionals can find these ideas very useful when working with the religious and spiritual populations in their communities. We can, for example, use the convergences as topics for community intergroup dialogues. We can also introduce these topics into the conferences, workshops, trainings, and other kinds of classes we offer so that people from all walks of life can be introduced to and think about these key emerging climate crisis issues.

TABLE 15.2 Seven Convergences of Major World Religions Regarding Climate Crisis

Seven Convergences	Brief Descriptions
First	Embrace a holistic view of nature as a sacred and divine creation.
Second	See humans as stewards of nature rather than owners.
Third	Accept scientific consensus of the etiologies, magnitude, and effects of climate crisis.
Fourth	Accept need to reduce greenhouse gases dramatically and increase renewables.
Fifth	Move away from limitless consumption and economic growth to sustainable practices that restore our own morality and spirituality.
Sixth	Commit to global social and climate justice.
Seventh	Redistribute economic power, especially in the energy sector.

Source: Summary from article by Chaplin, 2016.

The Green Convergence of Science, Religion, and Spirituality

In addition to the green convergence of religion and spirituality, perhaps there is also now a growing opportunity for a convergence of science, religion, and spirituality that can help support the climate solutions on which we need to focus. For example, a popular online series of conversations (Hoffman & Spira, 2024) between academic Donald Hoffman and spiritual teacher Rupert Spira suggests possible areas of agreement using different ways of knowing concerning the nature of matter and consciousness. Hoffman is a professor of cognitive sciences at the University of California, Irvine, with joint appointments in computer science, philosophy, and logic and philosophy of science, and Spira is an author, spiritual teacher, philosopher, and potter.

Hoffman has used math and science tools in his efforts to show how our current space-time theory may not be fundamental, as Einstein believed, but in reality secondary to human experience. Instead, he argues that it is Consciousness itself that actually gives rise to reality. Hoffman seeks to develop scientific theories that support the "perennial philosophy," or universal truths, that all the world's wisdom, religious, and spiritual traditions share. He hopes that scientists can move from the current predominant physicalist view that Consciousness is a product of the brain to a view in which Consciousness is seen as existing beyond space-time. He cites, for example, how the amplituhedron, a geometric form used to calculate quantum particle interactions, exists outside of space-time.

Spira sees Consciousness as the ultimate reality that is difficult to define but that in fact defines everything else. He sees the universality of Consciousness as a truth that points humanity toward the interconnection we have with everything in the Universe and describes our human experience as a part of a larger Consciousness. Consciousness is universal, and only Consciousness can know itself. He suggests that we think of the human mind as a virtual reality headset and imagine that the experience of space-time is created when Consciousness wears that headset and employs human thinking and perceiving.

Is there, as Spira and Campbell suggest, a fundamental Consciousness that is primary to everything in the Universe, including ourselves? And if they (and the many others who agree) are correct, could humanity embrace the implied connection we have with everything else in existence?

A recent article in *New Scientist* describes scientists' current efforts to test whether the quantum effects they believe exists in all matter throughout the Universe can actually cause consciousness. Anesthesiologist Stuart Hameroff and Nobel Prize–winning physicist Roger Penrose have proposed that consciousness has its origins in quantum phenomena in our brains. Now scientists are beginning to test their theories by studying tubulin proteins, the microtubules they form, and the possibility that these microtubules inside brain neurons can create gravitationally induced wave functions to collapse into consciousness. Although much more research is necessary, "there are tantalizing hints that quantum effects do exist inside the brain" (Musser, 2024, p. 35). Perhaps there may be further exciting discoveries about the "quantum mind" in our lifetime.

Additionally, perhaps, as helping professionals, we may need to also understand that even if scientists cannot yet prove that human consciousness connects us with the Consciousness that may well exist across the Universe, we may still be able to show people the benefits of living as if it was true. In other words, when we live *as if* we are connected with everything, it profoundly changes out attitude toward other people, other living things, and the ecosystems that keep us alive.

Many leaders in our spiritual circles and religious communities are calling for us to actually do just that: to begin to live more in connection with one another and the world we exist in. As Wayne Boulton (1990) wrote after reflecting on Fox's work:

> From the Mediterranean to Alaska to the Soviet Union to the California coast, we encounter news of ecological disaster. ... In this global crisis, Fox argues, political programs and voluntary activity will not be enough. A spiritual response is also required—the enormous resources of our religious heritage. The earth will continue to bestow its blessings of soil, forest and rain, but are we responding as we should—with gratitude, restraint, appropriate reverence and the proper rites? (p. 1)

Consciousness-Based Climate Solutions

The range of religious-spiritual interventions we can use in our climate-sensitive practice can be arranged by using the seven-paradigm inclusive practice framework we have studied in our text. In other words, spiritually oriented practice can include work with the past (psychodynamic) with beliefs and actions (CBT) with here-and-now awareness of emotions (experiential) with consciousness (transpersonal) with the body (body-mind-spirit) with the environment (eco-body-mind-spirit) and with faith-based activism (case management).

This framework is summarized in Table 15.3. (For more details on these strategies, see Derezotes, 2006.)

TABLE 15.3 Spiritual-Religious Paradigms of Practice, With Goals and Examples

Religious-Spiritual Practice Paradigms	Paradigm Goals	Specific Practice Examples
Psychodynamic (the past)	Become conscious of the religious-spiritual conditioning I have had in the past about other people, other living things, and ecosystems.	What was I taught about how God wants us care for and use Earth? How do I think now about my responsibility toward Earth?
Cognitive behavioral (beliefs and actions)	Become conscious of the religious-spiritual beliefs and behaviors I now have and now want to have about other people, other living things, and ecosystems.	What do I believe are humanity's rights and responsibilities regarding nature? What is my life purpose in relationship to my stewardship of Earth?
Experiential (emotions)	Become conscious of the religious and spiritual emotions and feelings I have in the here and now about other people, other living things, and ecosystems.	When do I feel like wildlife or other parts of nature are sacred to me? What does Mother Earth give to me, and what do I have a responsibility to give back?
Transpersonal (consciousness)	Become conscious of the nature of my own mind and of consciousness itself.	Where does my consciousness seem to come from? What is my relationship with my mind?

Body-mind-spirit connection (body)	Become conscious of my body-mind-spirit connection.	How do I identify with my body? When do I feel my body is sacred?
Eco-body-mind-spirit connection (environment)	Become conscious of my body-mind-spirit-environment connection.	How connected am I with the Universe? What kind of connection do I want to have with the Universe?
Religious-spiritual case management (faith-based activism)	Become engaged in faith-based climate activism.	What do I have to offer in my service to the world? What kind of activism would I enjoy the most?

What are some additional religious-spiritual climate solutions that we could introduce in this chapter? We will now review briefly some climate solutions through the lenses of consciousness and connection.

Addressing Spiritual Poverty

There are many definitions of *spiritual poverty* in the literature, most of which address the subject from a religious perspective. For example, the Billy Graham Evangelistic Association (2024) explains that to be:

> poor in spirit is to be aware of our spiritual poverty. No one is more pathetic than he who is in great need and not aware of it. The body can be strengthened by food and water, but the soul cannot not be nourished apart from the Spirit of the living God. The soul, created in the image of God, cannot be fully satisfied until it knows God in the proper way.

In contrast, Helminiak (2020) offered a social-scientific perspective on spiritual poverty in which he conceptualized the idea as a loss of spiritual sensitivity and as a psychological disorder. Both religious and psychology-based articles express concern that spiritual poverty is a growing, widespread, and serious issue.

We can also define spiritual poverty as a loss of consciousness of the interconnected nature of ourselves and our world. Perhaps there is a growing yearning in our culture for a more conscious existence as humanity becomes more aware of the reality of our climate crisis. Such a desire might very well be critical in informing our efforts to insure our collective well-being and survival.

Mysticism and Psychedelic Medicines

Mysticism is a form of religious and spiritual practice that builds upon our firsthand experience of the divine and sacred. The concept of "mystical experience" emphasizes the experiential aspect of connection with the world. Mystical experiences can often be associated with the quest by people to experiences a transformation in consciousness and connection (McGinn, 2006).

According to Fox:

> The mystics often talk about the dark night of the soul as a school. ... This is the next step of evolution. ... Either we're going to evolve rapidly and undergo deep transformation, or we will go

> the way of all our other brother and sister hominids such as the Neanderthal and the Denisovans and all these other extinct species we are finding evidence of today. (Harvey, 2023)

One of the increasingly popular forms of mystical experiences sought by people in our culture is the use of psychedelic medicines to enhance our connection with ourselves and the Universe. There are currently an estimated 30 million people who use psychedelics in the United States for many purposes—some for recreation, some for relief of mental disorders, and some for the expansion of consciousness. As more states consider legalizing the use of psychedelic medicines, researchers continue to explore the benefits and risks of their use. Increasing numbers of people are now experimenting with the medicine and talking and writing about their experiences (see, for example, Pollan, 2019).

Psychedelic medicines have been used by ancient people for thousands of years, often during sacred ceremonies. The *UC Berkeley News* Changemaker series highlights "innovative members of the campus community engaged in work and research that tackles society's most pressing issues." In a recent article, the series highlights the work of Yuria Celidwen, who encourages scientists studying the use of hallucinogenic medicines to consider the Indigenous spirit medicines in their research. She states, "These medicines are not about the human mind alone; they reveal Spirit, the very animating principle of Life." (Natividad, 2023, p. 1).

Celidwen was born into an Indigenous Nahua and Mayan descent "family of mystics, healers, poets and explorers from the highlands of Chiapas, Mexico." She explained that she:

> grew up with one foot in the wilderness and another in the magical realism of Indigenous culture. My Elders' songs and stories enthralled my childhood. They enhanced my mythic imagination and emotional intuition, which became the fertile soil where the seeds of kindness, play and wonder dig their roots. (Natividad, 2023, p. 1)

Like any other spiritual path, the use of psychedelic medicines is not for everyone. These substances have both potential benefits and risks, and most are still considered controlled substances (Grob & Grigsby, 2021). Professional helpers need to educate ourselves about these benefits and risks, since we are increasingly likely to encounter community members who want to incorporate hallucinogenic medicines into their psychological, religious, and spiritual work. For some people, psychedelic experiences may open the door to new religious and spiritual insights about the interconnectedness between themselves and the world. However, these medicines are no substitute for the often challenging work of conscious meditation, dialogue, and activism that is also necessary for us to effectively and collectively address our climate crisis.

Bridging Cultural Divides Through Science, Religion, and Spirituality

In Section 1, we examined how the human ego and associated left-brain thinking may be the biggest obstacle to human cooperation for climate solutions. Remember that ego can be understood as the identification with form, especially with our mental forms and beliefs. Our political beliefs currently are interfering with our progress in addressing our climate crisis. A recent article in *USA Today*, for example, explained that more than 15% of U.S. counties have now effectively blocked the construction of new wind and/or solar installations (Weise et al., 2024).

Social scientists have discovered that we humans tend to approve of policies that come from members of their own in-group and dislike policies formed by out-group members. In addition, we also know that our moral judgments are usually more lenient with the behaviors of in-group members and more severe with those of out-group members. Such research also suggests that positive intergroup contact and voting reform can help bridge the differences that now divide us. Voting reform can include the ability to elect members from more than two parties at elections and to vote directly for policies rather than for parties (De-Wit et al., 2019).

As we currently progress through another U.S. presidential election year, it appears that the temptation to use climate denial for political gain continues to be too strong for many politicians to resist. We have explored these issues in previous chapters in this text from a variety of perspectives. In this last chapter, we can also briefly consider how the emerging convergence of science, religion, and spirituality we have been considering may also help us bridge the differences that currently divide us and learn to cooperate in addressing the climate crisis.

On a grassroots level, we helping professionals can offer opportunities for local climate crisis intergroup dialogues that include people with various opposing scientific, religious, spiritual, political, and other perspectives. Such conversations could help communities find mutual civility, establish common ground, and collectively resist the attempts by some politicians and some corporate leaders to divide us.

Protection of Extraterrestrial Anthromes and Biomes

As we near the end of this text, we can note that the human race is in yet another space race. There are daily reminders of this competition as we read about rocket launches, new satellites, and plans for landings on the moon, Mars, and nearby asteroids. Inevitably, humans will enter new unexplored extraterrestrial wildernesses and thereby risk causing harm to those environments.

The issues are complex and probably still largely unknown, but there are precautions we can take to reduce risk. For example, one issue that humans will need to study in the coming decades is that of *terraformation*. Terraforming involves the modification of the ecology of a planet or another extraterrestrial body so that it is habitable for human life. Carl Sagan wrote about the possibility of humans terraforming the planets Venus and Mars so that we could live on these neighboring planets without having to wear cumbersome spacesuits outside and perhaps also grow crops and raise farm animals outdoors (Sagan et al., 2013).

However, the terraformation of a new planet could potentially destroy any and all alien life forms that previously existed on a planet. In addition, in traveling back and forth, humans could accidently bring back to Earth alien life forms that could then infect our home planet. These risks raise ethical and practical issues that will need to be addressed (Nesvold, 2024).

Earth Day

Today happens to be Earth Day here on 2024 planet Earth. Earth Day is an annual celebration that enhances appreciation of the beauty of our planet, recognizes the efforts of the environmental movement, and fosters consciousness of our climate crisis. It is celebrated on April 22 in the United States and many other countries and also sometimes on the Spring Equinox in other locations in the world. A quick look at the current EarthDay.org website (https://www.earthday.org/) reveals such headlines as a call for a major reduction in plastic production, a description of climate education, and an article on "The Great Global Cleanup."

I believe that humanity is learning the lessons necessary to successfully address our climate crisis. Our transformations in consciousness and evolving sense of connection will not only inform our ability to successfully address the climate crisis but also serve us in addressing the other major interrelated challenges of our era, including xenophobia, economic and political inequality, preparations for war and terror, risks of future pandemics, and human overpopulation.

My faith-based activism tells me that our global environmental movement will thus not only contribute to the survival and well-being of humanity, other living things, and our ecosystems but also serve to inspire many future human generations who no doubt will have their own opportunities to enjoy and daunting challenges to overcome.

May every day be Earth Day!

QUESTIONS FOR REFLECTION

1. What are your hopes and fears about the convergence of science, religion, and spirituality that we theorized in this chapter?
2. Please watch the videos of the conversations between Hoffman and Spira. What are your reactions?
3. Now that you have read the text, what is your overall reaction? How do you think and feel now about the climate crisis? What is important for you personally to do next?

REFERENCES

Alper, B. A., Rotolo, M., Tevington, P., Nortey, J., & Kallo, A. (2023). *Who are "spiritual but not religious" Americans?* Pew Research Center, December 7, 2023. https://www.pewresearch.org/religion/2023/12/07/who-are-spiritual-but-not-religious-americans/

Billy Graham Evangelistic Association staff. (2024). *Q: I heard a sermon that stated that just because we do good things doesn't make us fit for Heaven and that we are in "spiritual poverty," but what does that mean?* Answers, January 19, 2024. https://billygraham.org/answer/what-does-spiritual-poverty-mean/

Boulton, W. G. (1990). *The thoroughly modern mysticism of Matthew Fox.* Reprinted by Religion Online. https://www.religion-online.org/article/the-thoroughly-modern-mysticism-of-matthew-fox/

Chaplin, J. (2016). The global greening of religion. *Palgrave Communications, 2*, 16047. https://doi.org/10.1057/palcomms.2016.47

Dasgupta, P., & Ramanathan, V. (2014). Pursuit of the common good. *Science*, 345(6203), 1457–1458.

De-Wit, L., van fer Linden, S., & Brick, C. (2019). What are the solutions to political polarization? *Greater Good Magazine*, July 2, 2019. https://greatergood.berkeley.edu/article/item/what_are_the_solutions_to_political_polarization

Derezotes, D. S. (2006). *Spiritually oriented social work practice.* Pearson.

Ehrlich, P. (1988). Chapter 2: The loss of biodiversity: Causes and consequences. In E. O. Wilson & F. M. Peter (Eds.), *Biodiversity.* National Academies Press.

Fox, M. (1991). *Creation spirituality: Liberating gifts for the people of the Earth.* HarperOne.

Fox, M. (2012). *Hildegard of Bingen: A saint for our times: Unleashing her power in the 21st century.* Namaste Publishing.

Fox, M. (2024). *Welcome from Matthew Fox.* https://www.matthewfox.org/matthew-fox

Goodreads. (2024). *Matthew Fox quotes.* https://www.goodreads.com/quotes/8815080-the- doctrine-of-the-incarnation-is-itself-an-invitation-to

Gottlieb, R. S. (2006). Introduction: Religion and ecology—What is the connection and why does it matter? In *The Oxford Handbook on Religion and Ecology* (pp. 1–21). Oxford University Press.

Grob, C. S.. & Grigsby, J. (Eds.; 2021). *Handbook of medical hallucinogens.* Guilford.

Harvey, A. (2023). *Interview with Matthew Fox: Essential writings on creation spirituality.* Tikkun.org, May 19, 2023. https://www.tikkun.org/interview-with-matthew-fox-on-essential-writings-on-creation-spirituality/

Helminiak, D. A. (2020). Material and spiritual poverty: A postmodern psychological perspective on a perennial problem. *Journal of Religious Health*, *59*, 1458–1480. https://doi.org/10.1007/s10943-019-00873-z

Hoffman, D., & Spira, R. (2024). *The convergence of science and spirituality.* YouTube. Part 1: https://www.youtube.com/watch?app=desktop&v=rafVevceWgs; Part 2: https://www.youtube.com/watch?v=ZV-DMJpvRXc

Lennon, J., & P. McCartney. (1970). "Two of Us." Azlyrics. https://www.azlyrics.com/lyrics/beatles/twoofus.html

Lipka, M., & Gecewicz, C. (2017). *More Americans now say they're spiritual but not religious.* Pew Research Center, September 6, 2017. https://www.pewresearch.org/short-reads/2017/09/06/more-americans-now-say-theyre-spiritual-but-not-religious/

McGinn, B. (2006). *The essential writings of Christian mysticism.* Modern Library.

Musser, G. (2024). The quantum mind. *New Scientist*, 32–35.

Natividad, I. (2023). *Why Indigenous "Spirit medicine" principles must be a priority in psychedelic research.* UC Berkeley News, May 3, 2023. https://news.berkeley.edu/2023/05/03/why-indigenous-spirit-medicine-principles-must-be-a-priority-in-psychedelic-research/

Nesvold, E. (2024). What we should think about before terraforming alien worlds. *Popular Science*, February 24, 2024. https://www.popsci.com/science/why-terraform-alien-worlds/

Pollan, M. (2019). *How to change your mind: The new science of psychedelics.* Penguin Books.

Sagan, C., Druyan, A., & Tyson, N. D. (2013). *Cosmos.* Ballantine Books.

Weise, E., Beard, S. J., Bhat, S., Padilla, R., Procell, C. P., & Zaiets, K. (2024). U.S. counties are blocking the future of renewable energy: These maps, graphics show how. *USA Today*, February 4, 2024. https://www.usatoday.com/story/graphics/2024/02/04/us-renewable-energy-grid-maps-graphics/72042529007/

Credit

Index

www.ingramcontent.com/pod-product-compliance
Ingram Content Group UK Ltd.
Pitfield, Milton Keynes, MK11 3LW, UK
UKHW050141280726
14058UKWH00006B/770

9 798823 330312